Research Notes in Mathematics

D1768834

Main Editors
A. Jeffrey, University of Newcastle-upon-Tyne
R. G. Douglas, State University of New York at Stony Brook

Editorial Board
F. F. Bonsall, University of Edinburgh
H. Brezis, Université de Paris
G. Fichera, Università di Roma
R. P. Gilbert, University of Delaware
K. Kirchgässner, Universität Stuttgart
R. E. Meyer, University of Wisconsin-Madison
J. Nitsche, Universität Freiburg
L. E. Payne, Cornell University
G. F. Roach, University of Strathclyde
I. N. Stewart, University of Warwick
S. J. Taylor, University of Virginia

Submission of proposals for consideration
Suggestions for publication, in the form of outlines and representative samples, are invited by the editorial board for assessment. Intending authors should contact either the main editor or another member of the editorial board, citing the relevant AMS subject classifications. Refereeing is by members of the board and other mathematical authorities in the topic concerned, located throughout the world.

Preparation of accepted manuscripts
On acceptance of a proposal, the publisher will supply full instructions for the preparation of manuscripts in a form suitable for direct photo-lithographic reproduction. Specially printed grid sheets are provided and a contribution is offered by the publisher towards the cost of typing.

Illustrations should be prepared by the authors, ready for direct reproduction without further improvement. The use of hand-drawn symbols should be avoided wherever possible, in order to maintain maximum clarity of the text.

The publisher will be pleased to give any guidance necessary during the preparation of a typescript, and will be happy to answer any queries.

Important note
In order to avoid later retyping, intending authors are strongly urged not to begin final preparation of a typescript before receiving the publisher's guidelines and special paper. In this way it is hoped to preserve the uniform appearance of the series.

Titles in this series

1 Improperly posed boundary value problems
A Carasso and A P Stone

2 Lie algebras generated by finite dimensional ideals
I N Stewart

3 Bifurcation problems in nonlinear elasticity
R W Dickey

4 Partial differential equations in the complex domain
D L Colton

5 Quasilinear hyperbolic systems and waves
A Jeffrey

6 Solution of boundary value problems by the method of integral operators
D L Colton

7 Taylor expansions and catastrophes
T Poston and I N Stewart

8 Function theoretic methods in differential equations
R P Gilbert and R J Weinacht

9 Differential topology with a view to applications
D R J Chillingworth

10 Characteristic classes of foliations
H V Pittie

11 Stochastic integration and generalized martingales
A U Kussmaul

12 Zeta-functions: An introduction to algebraic geometry
A D Thomas

13 Explicit *a priori* inequalities with applications to boundary value problems
V G Sigillito

14 Nonlinear diffusion
W E Fitzgibbon III and H F Walker

15 Unsolved problems concerning lattice points
J Hammer

16 Edge-colourings of graphs
S Fiorini and R J Wilson

17 Nonlinear analysis and mechanics: Heriot-Watt Symposium Volume I
R J Knops

18 Actions of fine abelian groups
C Kosniowski

19 Closed graph theorems and webbed spaces
M De Wilde

20 Singular perturbation techniques applied to integro-differential equations
H Grabmüller

21 Retarded functional differential equations: A global point of view
S E A Mohammed

22 Multiparameter spectral theory in Hilbert space
B D Sleeman

24 Mathematical modelling techniques
R Aris

25 Singular points of smooth mappings
C G Gibson

26 Nonlinear evolution equations solvable by the spectral transform
F Calogero

27 Nonlinear analysis and mechanics: Heriot-Watt Symposium Volume II
R J Knops

28 Constructive functional analysis
D S Bridges

29 Elongational flows: Aspects of the behaviour of model elasticoviscous fluids
C J S Petrie

30 Nonlinear analysis and mechanics: Heriot-Watt Symposium Volume III
R J Knops

31 Fractional calculus and integral transforms of generalized functions
A C McBride

32 Complex manifold techniques in theoretical physics
D E Lerner and P D Sommers

33 Hilbert's third problem: scissors congruence
C-H Sah

34 Graph theory and combinatorics
R J Wilson

35 The Tricomi equation with applications to the theory of plane transonic flow
A R Manwell

36 Abstract differential equations
S D Zaidman

37 Advances in twistor theory
L P Hughston and R S Ward

38 Operator theory and functional analysis
I Erdelyi

39 Nonlinear analysis and mechanics: Heriot-Watt Symposium Volume IV
R J Knops

40 Singular systems of differential equations
S L Campbell

41 N-dimensional crystallography
R L E Schwarzenberger

42 Nonlinear partial differential equations in physical problems
D Graffi

43 Shifts and periodicity for right invertible operators
D Przeworska-Rolewicz

44 Rings with chain conditions
A W Chatters and C R Hajarnavis

45 Moduli, deformations and classifications of compact complex manifolds
D Sundararaman

46 Nonlinear problems of analysis in geometry and mechanics
M Atteia, D Bancel and I Gumowski

47 Algorithmic methods in optimal control
W A Gruver and E Sachs

48 Abstract Cauchy problems and functional differential equations
F Kappel and W Schappacher

49 Sequence spaces
W H Ruckle

50 Recent contributions to nonlinear partial differential equations
H Berestycki and H Brezis

51 Subnormal operators
J B Conway

52 Wave propagation in viscoelastic media
F Mainardi

53 Nonlinear partial differential equations and their applications: Collège de France Seminar. Volume I
H Brezis and J L Lions

54 Geometry of Coxeter groups
H Hiller

55 Cusps of Gauss mappings
T Banchoff, T Gaffney and C McCrory

56 An approach to algebraic K-theory
A J Berrick
57 Convex analysis and optimization
J-P Aubin and R B Vintner
58 Convex analysis with applications in the differentiation of convex functions
J R Giles
59 Weak and variational methods for moving boundary problems
C M Elliott and J R Ockendon
60 Nonlinear partial differential equations and their applications: Collège de France Seminar. Volume II
H Brezis and J L Lions
61 Singular systems of differential equations II
S L Campbell
62 Rates of convergence in the central limit theorem
Peter Hall
63 Solution of differential equations by means of one-parameter groups
J M Hill
64 Hankel operators on Hilbert space
S C Power
65 Schrödinger-type operators with continuous spectra
M S P Eastham and H Kalf
66 Recent applications of generalized inverses
S L Campbell
67 Riesz and Fredholm theory in Banach algebra
B A Barnes, G J Murphy, M R F Smyth and T T West
68 Evolution equations and their applications
F Kappel and W Schappacher
69 Generalized solutions of Hamilton-Jacobi equations
P L Lions
70 Nonlinear partial differential equations and their applications: Collège de France Seminar. Volume III
H Brezis and J L Lions
71 Spectral theory and wave operators for the Schrödinger equation
A M Berthier
72 Approximation of Hilbert space operators I
D A Herrero
73 Vector valued Nevanlinna Theory
H J W Ziegler
74 Instability, nonexistence and weighted energy methods in fluid dynamics and related theories
B Straughan
75 Local bifurcation and symmetry
A Vanderbauwhede
76 Clifford analysis
F Brackx, R Delanghe and F Sommen
77 Nonlinear equivalence, reduction of PDEs to ODEs and fast convergent numerical methods
E E Rosinger
78 Free boundary problems, theory and applications. Volume I
A Fasano and M Primicerio
79 Free boundary problems, theory and applications. Volume II
A Fasano and M Primicerio
80 Symplectic geometry
A Crumeyrolle and J Grifone
81 An algorithmic analysis of a communication model with retransmission of flawed messages
D M Lucantoni
82 Geometric games and their applications
W H Ruckle
83 Additive groups of rings
S Feigelstock
84 Nonlinear partial differential equations and their applications: Collège de France Seminar. Volume IV
H Brezis and J L Lions
85 Multiplicative functionals on topological algebras
T Husain
86 Hamilton-Jacobi equations in Hilbert spaces
V Barbu and G Da Prato
87 Harmonic maps with symmetry, harmonic morphisms and deformations of metrics
P Baird
88 Similarity solutions of nonlinear partial differential equations
L Dresner
89 Contributions to nonlinear partial differential equations
C Bardos, A Damlamian, J I Díaz and J Hernández
90 Banach and Hilbert spaces of vector-valued functions
J Burbea and P Masani
91 Control and observation of neutral systems
D Salamon
92 Banach bundles, Banach modules and automorphisms of C*-algebras
M J Dupré and R M Gillette
93 Nonlinear partial differential equations and their applications: Collège de France Seminar. Volume V
H Brezis and J L Lions
94 Computer algebra in applied mathematics: an introduction to MACSYMA
R H Rand
95 Advances in nonlinear waves. Volume I
L Debnath
96 FC-groups
M J Tomkinson
97 Topics in relaxation and ellipsoidal methods
M Akgül
98 Analogue of the group algebra for topological semigroups
H Dzinotyiweyi
99 Stochastic functional differential equations
S E A Mohammed
100 Optimal control of variational inequalities
V Barbu
101 Partial differential equations and dynamical systems
W E Fitzgibbon III
102 Approximation of Hilbert space operators. Volume II
C Apostol, L A Fialkow, D A Herrero and D Voiculescu
103 Nondiscrete induction and iterative processes
V Ptak and F-A Potra
104 Analytic functions – growth aspects
O P Juneja and G P Kapoor
105 Theory of Tikhonov regularization for Fredholm equations of the first kind
C W Groetsch

106 Nonlinear partial differential equations and free boundaries
J I Díaz

107 Variational convergences for functions and operators
H Attouch

108 A layering method for viscous, incompressible L_p flows occupying R^n
A Douglis and E B Fabes

The theory of Tikhonov regularization for Fredholm equations of the first kind

C W Groetsch

University of Cincinnati

The theory of Tikhonov regularization for Fredholm equations of the first kind

Pitman Advanced Publishing Program

BOSTON · LONDON · MELBOURNE

PITMAN PUBLISHING LIMITED
128 Long Acre, London WC2E 9AN

PITMAN PUBLISHING INC
1020 Plain Street, Marshfield, Massachusetts 02050

Associated Companies
Pitman Publishing Pty Ltd, Melbourne
Pitman Publishing New Zealand Ltd, Wellington
Copp Clark Pitman, Toronto

© C W Groetsch 1984

First published 1984

AMS Subject Classifications: (main) 65J05, 45L05, 65R05
(subsidiary) 45B05, 47A50

Library of Congress Cataloging in Publication Data

Groetsch, C. W.
The theory of Tikhonov regularization for Fredholm equations of the first kind.

Bibliography: p.
Includes index.
1. Fredholm equations—Numerical solutions. I. Title.
II. Title: Tikhonov regularization for Fredholm equations of the first kind.
QA431.G77 1984 519.4 83-25002
ISBN 0-273-08642-1

British Library Cataloguing in Publication Data

Groetsch, C. W.
The theory of Tikhonov regularization for Fredholm equations of the first kind.—(Research notes in mathematics; 105)
1. Banach algebras 2. Fredholm operators
I. Title II. Series
512'.55 QA326

ISBN 0-273-08642-1

All rights reserved. No part of this publication may be reproduced, stored in a retrieval system, or transmitted, in any form or by any means, electronic, mechanical, photocopying, recording and/or otherwise, without the prior written permission of the publishers. This book may not be lent, resold, hired out or otherwise disposed of by way of trade in any form of binding or cover other than that in which it is published, without the prior consent of the publishers.

Reproduced and printed by photolithography
in Great Britain by Biddles Ltd, Guildford

To my father and the memory of my mother.

Every restriction corresponds to a law of nature,

a regularization of the universe.

Carl Sagan

Preface

It has been two decades since the publication of Tikhonov's groundbreaking paper on the method of regularization for numerical solution of Fredholm integral equations of the first kind. The ensuing years have seen an intensive development of the theory of the method as well as its increasing application to difficult technical problems. A coherent and self-contained treatment of the general theory of Tikhonov's method for compact operators in Hilbert space would seem to be a timely and worthwhile undertaking. These notes represent our own modest attempt at such a treatment.

Our development is approximative rather than numerical, that is, the approximations themselves lie in the same Hilbert space as the solution and questions of convergence, rates, etc., are addressed in the general Hilbert space context. Although most of the results apply to more general operators, we limit our attention to compact operators, which results in considerable simplification, because our prime motivating example is a Fredholm integral equation of the first kind.

A reference in the text of the form "(a.b.c)" refers to theorem number "c" in section "b" of chapter "a." Equations and other important displayed items are numbered consecutively within each section. The end of a proof is indicated by the symbol "#."

These notes comprise the text of a course of lectures on Tikhonov regularization which I gave at the University of Kaiserslautern in the spring of 1983. I wish to thank Professor Eberhard Schock for inviting me to lecture in Germany.

C.W.G.
Kaiserslautern
June 1983

Contents

1 Introduction and preliminaries

In this chapter we introduce Fredholm equations of the first kind and the peculiar problems associated with their solution. This leads to a general discussion of ill-posed problems for operator equations of the first kind. Finally, the operator theoretic foundation for the sequel is laid by a discussion of compact operators on Hilbert space, spectral theory, Picard's theorem and generalized inverses.

1.1. Equations of the first kind; ill-posed problems

By a Fredholm integral equation of the first kind is meant an equation of the form

$$\int_a^b k(x,s)u(s)ds = g(x), \tag{1}$$

where g is a given function (usually called the "data"), $k(\cdot,\cdot)$ is a given function (the "kernel" of the equation) and the solution u is an unknown function which is sought. Several observations concerning this equation come immediately to mind. The first is that the function g inherits some of the smoothness of the kernel k and therefore a solution may not exist if g is too roughly behaved. For example, if the kernel k is continuous and u is integrable, then it is easy to see that the function g defined by (1) is also continuous and hence if the given function g is not continuous, while the kernel is, then (1) can have no integrable solution. This is simply to say that the question of *existence* of solutions is not trivial and requires investigation.

Another point to consider is *uniqueness* of solutions. For example, if $k(x,s) = x \sin s$, then the function $u(x) = 1/2$ is a solution of

$$\int_0^\pi k(x,s)u(s)ds = x,$$

but so is each of the functions $u_n(s) = 1/2 + \sin ns$, $n = 1,2,3,\ldots$.

A more serious concern arises from the Riemann-Lebesgue lemma which states that if $k(\cdot,\cdot)$ is any square integrable kernel, then

$$\int_0^{\pi} k(x,s)\sin ns\, ds \to 0 \quad \text{as} \quad n \to \infty.$$

From this it follows that if u is a solution of (1) and A is arbitrarily large, then

$$\int_0^{\pi} k(x,s)(u(s) + A \sin ns) ds \to g(x) \quad \text{as} \quad n \to \infty.$$

Therefore for large values of n the slightly perturbed data

$$\tilde{g}(x) = g(x) + A \int_0^{\pi} k(x,s)\sin ns\, ds$$

corresponds to a solution $u(s) + A \sin ns$ which differs markedly from $u(s)$. To put it another way, in Fredholm equations of the first kind, solutions generally depend *discontinuously* upon the data.

As a tangible example of the discontinuous dependence of the solution on the data, consider the problem of solving the one-dimensional heat equation "backward in time." That is, we imagine a metal bar, which for convenience we take to extend over the interval $0 \leqslant x \leqslant \pi$, whose temperature at the point x and at time t is given by the function $U(x,t)$. Then, for an appropriate choice of units, U satisfies the equation

$$\frac{\partial U}{\partial t} = \frac{\partial^2 U}{\partial x^2}, \qquad 0 < x < \pi. \tag{2}$$

If we take the initial temperature distribution of the bar to be the function $u(x) := U(x,0)$ and impose some boundary conditions, say

$$U(0,t) = U(\pi,t) = 0, \tag{3}$$

then the familiar method of separation of variables gives the following relationship between the temperature distribution $g(x) := U(x,1)$ at time $t = 1$ and the initial temperature distribution u:

$$g(x) = \int_0^{\pi} k(x,s)u(s)ds \tag{4}$$

where

$$k(x,s) = \frac{2}{\pi} \sum_{n=1}^{\infty} e^{-n^2} \sin nx \quad \sin ns.$$

By solving the heat equation backward in time we mean solving (4) for the initial temperature distribution u, given the temperature distribution g at a later time. The discussion above shows that the problem of numerically determining $u \in L^2[0,\pi]$ for a given $g \in L^2[0,\pi]$ may be difficult indeed. Of course the reason for this is that the mapping $g \to u$ is discontinuous, while in any practical situation g is not known precisely but only a measured approximation is in hand. Therefore small errors in the measured g may lead to large errors in the solution u. We shall see in the next section that this discontinuity of the solution operator for equation (1) is an inevitable consequence of the compactness of the integral operator.

There are many other examples of practical problems which lead to Fredholm integral equations of the first kind. Most of these examples arise from problems in remote sensing, e.g. location of tumors by tomography, determination of atmospheric temperature profiles from telemetry data, seismic prospecting, system identification, etc. We refer the reader to [1,3,6-10,12-13] and the forthcoming conference proceedings [11] for details of some of these problems.

Let us now consider problem (1) in the abstract as an equation of the first kind,

$$Ku = g, \tag{5}$$

where K is a mapping from some topological space X into a topological space Y. At the beginning of this century J. Hadamard formalized the concept of *well-posedness* for such equations as follows: equation (5) is said to be well-posed if

(a) for each $g \in Y$, there is a solution $u \in X$ of (5);

(b) the solution u is unique in X;

(c) the dependence of u upon g is continuous.

It is evident from this definition that the well-posedness of (5) is intimately connected not only with the operator K but also with the spaces X and Y and the topologies they carry, that is, well-posedness is a property of the *triple* (K,X,Y). Clearly for (a) to hold we must have $Y = K(X)$, that is, the mapping $K : X \to Y$ must be surjective. Condition (b) is of course equivalent to injectivity of K. Therefore if (a) and (b) hold, the mapping is invertible; condition (c) is merely another way of saying that the inverse mapping is continuous.

An equation of the type (5) which is not well-posed is called *ill-posed* and a method for approximately solving an ill-posed problem is called a *regularization* method. We shall assume for now that we wish to regularize an equation which satisfies (a) and (b) (in the next section we will see how these restrictions may be relaxed). Two broad ideas for regularization come immediately to mind: namely, one could attempt to satisfy (c) by modifying the spaces or by modifying the operator.

A general result along the lines of this first idea is a classical lemma of Tikhonov which has become a standard exercise in topology courses. Specifically, if K is continuous and injective when restricted to a compact subset $\hat{X}$ of X, then of course K maps closed subsets of $\hat{X}$ into closed subsets of Y, that is, K^{-1} is continuous on $K(\hat{X})$ and the problem $(K,\hat{X},K(\hat{X}))$ is well-posed.

The imposition of compactness may in fact be a needlessly strong restriction. In the case of the heat conduction equation considered above, if we assume the existence of a temperature distribution $U(x,t_o)$ with $t_o < 0$ satisfying an *a priori* bound of the type

$$\| U(\cdot,t_o)\|^2_{L^2} \leq C,$$

then, assuming that U is sufficiently smooth, and setting

$$F(t) = \int_{t_o}^{\pi} (U(x,t))^2 dx,$$

it is not difficult to show, using (2) and (3), that

$$\frac{d^2}{dt^2} \ln F(t) \geq 0,$$

that is, F(t) is *logarithmically convex* ([9], [12], [2]). Since $\ln F(t)$ is convex and $F(t_o) \leq C$ it then follows that for $t_o \leq t \leq 1$,

$$\ln F(t) \leq \frac{1-t}{1-t_o} \ln C + \frac{t-t_o}{1-t_o} \ln F(1)$$

and hence

$$\| U(\cdot,0) \|^2_{L^2} \leqslant C^{\frac{1}{1-t_o}} \| U(\cdot,1) \|_{L^2}^{\frac{2t_o}{t_o-1}} .$$

By the linearity of (2) - (3) it follows that the dependence of the temperature $U(\cdot,0)$ on the later temperature $U(\cdot,1)$ is continuous in the L^2- sense .

In these notes we will be exclusively concerned with the type of regularization which results when the spaces are not disturbed but the operator is modified so as to produce an operator with a continuous inverse. For example, the operator in (4) related to the backward-in-time heat flow problem is self-adjoint and has no negative eigenvalues. Therefore, for any $\alpha > 0$, the integral equation of the *second kind*

$$g(x) = \int_0^{\pi} k(x,s)u(s)ds + \alpha u(x) \tag{6}$$

has a unique solution which depends continuously on g, i.e., is well-posed. Hence one might reasonably try to approximate a solution of (4) by the solution of (6) for small positive values of α. Our aim in the subsequent chapters is to develop the theory of a regularization method of this general type, *Tikhonov regularization,* for compact but not necessarily self-adjoint operators on Hilbert space. But first we will lay out some background on operator theory which will be needed in the sequel.

1.2. Linear operators in Hilbert space

Our purpose in this section is to establish some notation and provide a brief tutorial on some basic theorems of operator theory.

The main results of this section can be found in most books on functional analysis (e.g. [5]). We will use the symbol $(\cdot,\cdot)$ to denote the generic inner product in a Hilbert space and the symbol $\|\cdot\|$ to designate the associated norm. If S is a subset of a Hilbert space H, then $\overline{S}$ will denote the (norm) closure of S and $S^{\perp}$ will denote the orthogonal complement of S, i.e.,

$$S^{\perp} = \{y \in H: (x,y) = 0 \text{ for all } x \in S\}.$$

If T is a continuous linear operator from a Hilbert space H_1 into a Hilbert space H_2, then its *adjoint* will be denoted T^*. That is, $T^*: H_2 \to H_1$ is the continuous linear operator defined by

$$(Tx,y) = (x,T^*y)$$

for all $x \in H_1$ and $y \in H_2$. The *range*, $R(T)$, and *nullspace*, $N(T)$, of a linear operator with domain $\mathcal{D}(T)$ are defined by

$$R(T) = \{Tx : x \in \mathcal{D}(T)\}$$

and

$$N(T) = \{x \in \mathcal{D}(T) : Tx = 0\},$$

respectively. The following result connecting these concepts is fundamental.

THEOREM 1.2.1. <u>If</u> $T : H_1 \to H_2$ <u>is a continuous linear operator, then</u> $R(T)^\perp = N(T^*)$ <u>and</u> $N(T)^\perp = \overline{R(T^*)}$.

Since $T = T^{**}$ we find that $R(T^*)^\perp = N(T)$ and $N(T^*)^\perp = \overline{R(T)}$.

We recall that for linear operators continuity is equivalent to *boundedness*, that is, finiteness of the number

$$\|T\| = \sup_{\|x\|=1} \|Tx\|$$

called the *norm* of T. Moreover, if T is bounded, then

$$\|T\| = \|T^*\| = \|TT^*\|^{1/2}.$$

The *spectrum* of a linear operator $T : H \to H$ is the set of complex numbers $\sigma(T)$ defined by

$$\sigma(T) = \{\lambda \in \mathbb{C} : \ T - \lambda I \ \text{ has no bounded inverse}\}$$

where I is the identity operator on H. The *spectral radius* of T is the real number $|\sigma(T)|$ defined by

$$|\sigma(T)| = \sup\{|\lambda| : \lambda \in \sigma(T)\}.$$

If T is bounded, then $\sigma(T)$ is closed and $|\sigma(T)| \leq \|T\|$, therefore $\sigma(T)$ is a compact set.

An operator $T : H \to H$ is called *self-adjoint* if $T = T^*$. If T is a bounded self-adjoint linear operator, then $\sigma(T)$ is a nonempty set of real numbers. The next result is known as the *spectral radius formula*.

THEOREM 1.2.2. <u>If T : H → H is a bounded self-adjoint operator, then</u> $|\sigma(T)| = \|T\|$.

A complex number λ is called an *eigenvalue* of a linear operator $T : H \to H$ if $Tx = \lambda x$ for some nonzero vector x, called an *eigenvector*, associated with λ. Obviously every eigenvalue of T is a member of $\sigma(T)$. If T is self-adjoint then eigenvectors associated with distinct eigenvalues are orthogonal.

A linear operator $K : H_1 \to H_2$ is called *compact* if $\overline{K(B)}$ is compact for each (norm) bounded subset B of H_1. The general theory of compact operators evolved from the theory of integral operators of the form

$$Ku(x) = \int_a^b k(x,s)u(s)ds.$$

Indeed, if $k(\cdot,\cdot)$ is square integrable over $[c,d] \times [a,b]$, then it is a well known classical result that K is a compact operator from $L^2[a,b]$ into $L^2[c,d]$ (see, e.g., [5]).

A compact operator is continuous; in fact compact operators in Hilbert space may be characterized as linear operators which map weakly convergent sequences into strongly convergent sequences and hence compact operators are called *completely continuous* by some writers. From the definition it is clear that the composition of a compact operator with a bounded operator is also compact. It follows that a compact operator on a Hilbert space H is also compact on any Hilbert space which is continuously imbedded in H. In particular, the integral operator (1) generated by a square integrable kernel is also compact when considered as an operator on the Sobolev spaces $H_2^n[a,b]$.

Compact self-adjoint operators have a particularly simple spectrum: each nonzero member of the spectrum is an isolated point which is an eigenvalue of the operator. Moreover, for each nonzero eigenvalue λ of a compact self-adjoint operator K, the *eigenspace* associated with λ, that is, the set $N(K - \lambda I)$, is finite dimensional and the eigenvalues form a sequence $\lambda_1, \lambda_2, \ldots$, which (if infinite) converges to zero. If we repeat each eigenvalue in this list according to the dimension of its associated eigenspace we may form a sequence $w_1, w_2, \ldots$, of associated orthonormal eigenvectors. Finally, we are in a position to state the *spectral theorem* for a compact self-adjoint operator.

THEOREM 1.2.3. Suppose $K : H \to H$ is a compact self-adjoint linear operator with eigenvalues $\lambda_1, \lambda_2, \ldots$ (repeated according to the dimension of the associated eigenspace) and associated orthonormal eigenvectors $w_1, w_2, \ldots,$ then for any $x \in H$

$$Kx = \sum_n \lambda_n (x, w_n) w_n.$$

The sum in the theorem above may be finite or infinite depending upon whether K has finitely or infinitely many eigenvalues, respectively. If K has only finitely many eigenvalues it is said to be of *finite rank*, for in this case the spectral theorem shows that R(K) is finite dimensional and is spanned by the eigenspaces.

Of course the prototypes of compact operators on Hilbert space are the Fredholm integral operators on L^2 defined by an equation of the type (1). An operator of finite rank in this case corresponds to a *degenerate kernel*, that is, a kernel of the type

$$k(x,s) = \sum_{n=1}^{m} X_n(x) S_n(s).$$

Note that the kernel for the heat flow operator of the previous section is not degenerate and that $\{e^{-n^2} : n = 1,2,\ldots\}$ is the sequence of eigenvalues of this operator while

$$\{\sqrt{\tfrac{2}{\pi}} \sin ns : n = 1,2,\ldots\}$$

is a sequence of associated orthonormal eigenvectors.

The spectral theorem allows us to define functions of a compact self-adjoint operator K in the following way. Given a real-valued continuous function f on $\sigma(K)$ we define f(K) by

$$f(K)x = \sum_n f(\lambda_n)(x, w_n) w_n. \qquad (2)$$

It happens that the operator f(K) so defined is itself self-adjoint and compact and its spectrum is related to that of K by the *spectral mapping theorem*.

THEOREM 1.2.4. If K is compact and self-adjoint and f is a continuous real-valued function on $\sigma(K)$, then $\sigma(f(K)) = f(\sigma(K))$.

In view of (1.2.2) and (1.2.4) we have

$$\| f(K) \| = |\sigma(f(K))| = \sup\{|f(\lambda)| : \lambda \in \sigma(K)\}.$$

From this we obtain immediately the following useful approximation result.

THEOREM 1.2.5. *If, in addition to the hypotheses of* (1.2.4), $\{f_n\}$ *is a sequence of continuous real-valued functions which converge to* f *uniformly on* $\sigma(K)$, *then* $\| f_n(K) - f(K) \| \to 0$ *as* $n \to \infty$.

A special terminology involving "singular systems" is traditional in the study of Fredholm equations of the first kind. We now introduce this jargon in terms of the spectral theory outlined above. Suppose that

$$K : H_1 \to H_2$$

is a compact linear operator. Then

$$K^*K : H_1 \to H_1$$

is a compact self-adjoint linear operator and any eigenvalue β of K^*K satisfies

$$\beta = (\beta x, x) = (K^*Kx, x) = \| Kx \|^2 \geq 0$$

if x is an associated eigenvector of norm one. Therefore the nonzero eigenvalues of K^*K can be enumerated as $\lambda_1^2 \geq \lambda_2^2 \geq \ldots$. If we designate by $v_1, v_2, \ldots,$ an associated sequence of orthonormal eigenvectors and set

$$\mu_n = \lambda_n^{-1} \quad \text{and} \quad u_n = \mu_n K v_n,$$

then $\{u_n\}$ is an orthonormal sequence in H_2 and

$$\mu_n K^* u_n = v_n.$$

Moreover, it can be shown by use of the spectral theorem that $\{u_n\}$ is a complete orthonormal set for $\overline{R(K)} = N(K^*)^{\perp}$ and $\{v_n\}$ is a complete orthonormal set for $\overline{R(K^*)} = N(K)^{\perp}$. The sequence $\{u_n, v_n; \mu_n\}$ is called a *singular system* for K. The next result is known as *Picard's theorem* on the existence of solutions of first kind equations.

THEOREM 1.2.6. *Let* $K : H_1 \to H_2$ *be a compact linear operator with singular* system $\{u_n, v_n; \mu_n\}$. In order that the equation $Kz = g$ have a solution, it is necessary and sufficient that $g \in N(K^*)^{\perp}$ and

$$\sum_{n=1}^{\infty} \mu_n^2 \, |(g,u_n)|^2 < \infty.$$

The condition $g \in R(K)$ may be viewed as an abstract smoothness or regularity condition in the sense that g inherits some of the smoothness (with respect to the first variable) of the kernel. Picard's theorem reinterprets this regularity by requiring that the components $|(g,u_n)|$ decay quickly relative to the growth of the singular values (recall that $\mu_n \to \infty$ for nondegenerate kernels).

We see that the kernel of the integral operator for the heat equation ((4) of the previous section) is symmetric and nondegenerate. A singular system for the operator on $L^2[0,\pi]$ is given by

$$u_n(s) = v_n(s) = \sqrt{\frac{2}{\pi}} \sin ns; \qquad \mu_n = e^{n^2/2}.$$

Since $\{v_n\}$ is complete in $L^2[0,\pi]$ it follows that $N(K^*) = \{0\}$ and hence Picard's theorem says that a function $g \in L^2[0,\pi]$ is in the range of the operator if and only if its Fourier coefficients

$$g_n := (g, u_n)$$

decay rapidly enough to ensure that

$$\sum_{n=1}^{\infty} e^{n^2} |g_n|^2 < \infty.$$

In the sequel we will at times find it convenient to discuss a scale of conditions of the type $g \in R(K(K^*K)^{\nu})$, $\nu \geq 0$. Of course by $(K^*K)^{\nu}$ we mean the operator obtained via the representation (2) with $f(t) = t^{\nu}$. It follows easily from (1.2.4) that $\{u_n, v_n; \mu_n^{2\nu+1}\}$ is a singular system for $K(K^*K)^{\nu}$. Also, it is not difficult to show that

$$N(K^*) = N((K^*K)^{\nu}K^*) = N((K(K^*K)^{\nu})^*)$$

and hence we have immediately from Picard's theorem:

COROLLARY 1.2.7. *Suppose* $g \in N(K^*)^\perp$. *In order that* $g \in R(K(K^*K)^\nu)$, *it is necessary and sufficient that*

$$\sum_{n=1}^{\infty} \mu_n^{4\nu+2} |(g,u_n)|^2 < \infty.$$

1.3. Generalized inverses

Obviously the problem of solving

$$Ku = g \tag{1}$$

falls within the category of problems commonly referred to as "inversion." Indeed, to say that the triple (K,X,Y) is well-posed is precisely the same as saying that the inverse operator $K^{-1}: Y \to X$ exists and is continuous. In this section we introduce a somewhat more general notion of inversion for a bounded linear operator K from a Hilbert space H_1 into a Hilbert space H_2. Omitted proofs for theorems in this section can be found in [4].

In the first place, instead of insisting that equation (1) has a solution, we only require that there is some vector which is "solution-like" in the sense that

$$\| Ku - g\| = \inf \{\| Kx - g\| : x \in H_1\}.$$

Such a vector u is called a *least squares solution* of (1). The following easily proved theorem provides some equivalent characterizations of least squares solutions.

THEOREM 1.3.1. *Suppose* $K : H_1 \to H_2$ *is a bounded linear operator from a Hilbert space* H_1 *into a Hilbert space* H_2. *The following conditions are equivalent:*

(i) $\| Ku - g\| = \inf \{\| Kx - g\| : x \in H_1\}$,

(ii) $K^*Ku = K^*g$,

(iii) $Ku = Pg$,

where P is the orthogonal projection operator of H_2 *onto* $\overline{R(K)}$.

From (iii) we see that equation (1) has a least squares solution if and only if $Pg \in R(K)$, that is, if and only if g is a member of the dense subspace $R(K) + R(K)^\perp$ of H_2. Assuming this to be the case, each of the conditions shows that the set of least squares solutions is a closed convex set. This set of least squares solutions therefore has a unique element of smallest norm which is denoted $K^{\dagger}g$. The operator $K^{\dagger}$ defined on the dense subspace $\mathcal{D}(K^{\dagger}) = R(K) + R(K)^{\perp}$ in this way is called the Moore-Penrose *generalized inverse* of K. It is not difficult to show that $K^{\dagger}$ is a closed linear operator with $N(K^{\dagger}) = R(K)^{\perp}$ and $R(K^{\dagger}) = N(K)^{\perp}$. Also note that $K^{\dagger}g$ is the unique least squares solution of (1) lying in the subspace $N(K)^{\perp}$.

With this more general concept of solution, the existence and uniqueness questions are obviated for the triple $(K, N(K)^{\perp}, \mathcal{D}(K^{\dagger}))$. However, the stability problem may linger, as the following theorem shows.

THEOREM 1.3.2. $K^{\dagger}$ _is bounded if and only if_ R(K) _is closed_.

We are of course only interested in compact operators K and for such operators R(K) is closed only in trivial instances. Indeed, if K is compact and R(K) is closed then $K^{\dagger}$ is bounded and hence $KK^{\dagger}$ is compact. However $KK^{\dagger}$ is the identity operator when restricted to the Hilbert space R(K) and hence R(K) is finite dimensional. Therefore, for compact K, R(K) is closed if and only if it is finite dimensional. Hence we have

COROLLARY 1.3.3. _If_ $K : H_1 \to H_2$ _is a compact linear operator, then_ $K^{\dagger}$ _is bounded if and only if_ R(K) _is finite dimensional_.

We therefore see that the triple $(K, N(K)^{\perp}, \mathcal{D}(K^{\dagger}))$ is well-posed only in the relatively trivial case when K has finite rank. In the language of Fredholm integral equations this says that, even in the extended sense developed above, the first kind equation is ill-posed except when the kernel is degenerate. Of course for a degenerate kernel the problem of solution reduces to the relatively simple task of solving a finite system of linear algebraic equations.

The upshot of the discussion above is that we must inevitably devise some means of imposing stability when solving first kind Fredholm equations with nondegenerate kernels. The remaining chapters of these notes are mainly concerned with one method of doing so - the method of Tikhonov regularization. However, before leaving this chapter we wish to give an explicit representation of the Moore-Penrose generalized inverse of a compact operator.

THEOREM 1.3.4. _If_ $K : H_1 \to H_2$ _is a compact linear operator with singular system_ $\{u_n, v_n; \mu_n\}$ _and_ $g \in \mathcal{D}(K^{\dagger})$, _then_

$$K^{\dagger}g = \sum_{n=1}^{\infty} \mu_n (Pg, u_n) v_n = \sum_{n=1}^{\infty} \mu_n (g, u_n) v_n ,$$

<u>where</u> P <u>is the orthogonal projection operator of</u> H_2 <u>onto</u> $\overline{R(K)}$.

Proof We note that if $g \in \mathcal{D}(K^{\dagger})$, then $Pg \in R(K)$ and hence by (1.2.6)

$$\sum_{n=1}^{\infty} \mu_n^2 \, |(Pg, u_n)|^2 < \infty.$$

Also note that since $u_n \in \overline{R(K)}$,

$$(Pg, u_n) = (g, Pu_n) = (g, u_n).$$

We therefore see that the infinite series in the statement of the theorem converges in H_1. Also, since $\{v_n\} \subseteq N(K)^{\perp}$, it follows that the vector v defined by

$$v = \sum_{n=1}^{\infty} \mu_n (Pg, u_n) v_n$$

is also in $N(K)^{\perp}$. Moreover,

$$Kv = \sum_{n=1}^{\infty} \mu_n (Pg, u_n) K v_n = \sum_{n=1}^{\infty} (Pg, u_n) u_n = Pg.$$

Therefore by (1.3.1) (iii), v is a least squares solution which lies in $N(K)^{\perp}$, i.e., $v = K^{\dagger}g$. #

Finally we mention that in the sequel we will usually assume that $g \in R(K)$, rather than making the more general assumption that $g \in \mathcal{D}(K^{\dagger}) = R(K) + R(K)^{\perp}$. The reason for this is that g is generally not available but rather a corrupted version $\tilde{g} = g + \epsilon$, where ϵ is an error, is in hand. In assuming that $g \in \mathcal{D}(K^{\dagger})$ we have $Pg \in R(K)$ and hence we may in effect replace g by Pg, absorbing the component of g in $R(K)^{\perp}$ into the vector ϵ.

REFERENCES

1. Deuflhard, P. and Hairer, E., (Eds.) Numerical Treatment of Inverse Problems in Differential and Integral Equations, Birkhäuser, Boston, 1983.
2. Engl, H.W., Analysis und Numerik schlecht gestellter Probleme, Johannes Kepler Universität, Linz, Austria, 1980.
3. Gorenflo, R., (Ed.) Inkorrekt gestellte Probleme I, II, Conference Proceedings, Free University of Berlin, Berlin, 1977-78.
4. Groetsch, C.W., Generalized Inverses of Linear Operators: Representation and Approximation, Dekker, New York, 1977.
5. Groetsch, C.W., Elements of Applicable Functional Analysis, Dekker, New York, 1980.
6. Hämmerlin, G. and Hoffmann, K.H., (Eds.), Improperly Posed Problems and their Numerical Treatment, Birkhauser, Basel, 1983.
7. Hermann, G.T., Image Reconstruction from Projections: the Fundamentals of Computerized Tomography, Academic Press, New York, 1980.
8. Hermann, G.T. and Natterer, F., (Eds.) Mathematical Aspects of Computerized Tomography, Lecture Notes in Medical Informatics, vol.8, Springer-Verlag, New York and Heidelberg, 1981.
9. Knops, R.J., (Ed.) Symposium on Non-well-posed Problems and Logarithmic Convexity, Lecture Notes in Mathematics, vol.316, Springer-Verlag, New York and Heidelberg, 1973.
10. Lavrentiev, M.M., Some Improperly Posed Problems in Mathematical Physics, translated from the Russian, Springer-Verlag, New York, 1967.
11. Nashed, M.Z., (Ed.) Ill-posed Problems: Theory and Practice, Reidel, Dordrecht, to appear.
12. Payne, L.E., Improperly Posed Problems in Partial Differential Equations, CBMS Regional Conference Series in Applied Mathematics, Society for Industrial and Applied Mathematics, Philadelphia, 1975.
13. Tikhonov, A.N. and Arsenin, V.Y., Solutions of Ill-posed Problems, translated from the Russian, Wiley, New York, 1977.

2 A general regularization method

In this chapter we introduce a general class of regularization methods for Fredholm equations of the first kind which includes Tikhonov's method as a special instance. We will be concerned with the general question of convergence, in both the strong and weak sense, and with establishing general error bounds. The chapter ends with a discussion of some specific regularization methods. Most of the results of this chapter do not depend in an essential way on compactness; indeed many results may be easily generalized to bounded or even closed densely defined operators.

2.1. Convergence results

We now develop a family of regularization methods for the equation

$$Ku = g \tag{1}$$

where K is a compact linear operator from a Hilbert space H_1 into a Hilbert space H_2. To be more precise, we wish to construct some operators $R_\alpha : H_2 \to H_1$ which are *continuous* and approximate $K^\dagger$ in the sense that

$$R_\alpha g \to u := K^\dagger g \quad \text{as} \quad \alpha \to 0$$

for each $g \in \mathcal{D}(K^\dagger)$. In this section we assume that g is known exactly; the question of imprecise data is taken up in a later section. The main results of this section appear in a more general form in [7] and [4].

Throughout the sequel we will often denote the operator K^*K by $\tilde{K}$. We then see by (1.3.1) (ii) that $u := K^\dagger g$ satisfies

$$\tilde{K}u = K^*g.$$

Therefore if $\tilde{K}$ were invertible, then we would have $u = \tilde{K}^{-1}K^*g$. Even if $\tilde{K}$ is not invertible, we nevertheless hope to approximate u by a vector of the form

$$R_\alpha(\tilde{K})K^*g \qquad (\alpha > 0)$$

where $R_\alpha(t)$ is a continuous function on $\sigma(\tilde{K}) \subseteq [0, \|K\|^2]$ which in some sense approximates the function $f(t) = 1/t$ (see the discussion on spectral theory in Section 1.2). We note that if p is any polynomial, then

$$p(K^*K)K^* = K^*p(KK^*)$$

and therefore, in view of the Weierstrass approximation theorem, the same identity is valid for continuous functions on $\sigma(K^*K) = \sigma(KK^*)$. It therefore follows that if we denote, as we shall hereafter, the operator KK^* by $\hat{K}$, then

$$R_\alpha(\tilde{K})K^* = K^*R_\alpha(\hat{K}). \tag{2}$$

An important feature of the operators $R_\alpha(\tilde{K})$ is that they are continuous for each $\alpha > 0$ and we wish to impose conditions on the functions $R_\alpha(t)$ which will ensure that

$$\lim_{\alpha \to 0} R_\alpha(\tilde{K})K^*g = K^\dagger g$$

for each $g \in \mathcal{D}(K^\dagger)$. We shall assume that

$$R_\alpha(t) \to 1/t \quad \text{as} \quad \alpha \to 0 \text{ for each } t > 0 \tag{3}$$

and

$$|tR_\alpha(t)| \text{ is uniformly bounded.} \tag{4}$$

Under these conditions the following theorem [7] ensures convergence.

THEOREM 2.1.1. <u>Suppose that</u> $\{R_\alpha\}_{\alpha>0}$ <u>is a family of continuous real-valued functions on</u> $[0, \|K\|^2]$ <u>satisfying (3) and (4), then</u> $R_\alpha(\tilde{K})K^*g \to K^\dagger g$ <u>as</u> $\alpha \to 0$ <u>for each</u> $g \in \mathcal{D}(K^\dagger)$.

Proof By (2), $R_\alpha(\tilde{K})K^*g \in R(K^*)$ and therefore, in terms of a singular system $\{u_n, v_n; \mu_n\}$ for K, we have

$$\begin{aligned} R_\alpha(\tilde{K})K^*g &= \sum_{n=1}^{\infty} R_\alpha(\mu_n^{-2})\ (K^*g, v_n)v_n \\ &= \sum_{n=1}^{\infty} \mu_n^{-1} R_\alpha(\mu_n^{-2})\ (g,u_n)v_n \\ &= \sum_{n=1}^{\infty} \mu_n\, \mu_n^{-2} R_\alpha(\mu_n^{-2})\ (g,u_n)v_n. \end{aligned}$$

But by (3), (4) and the bounded convergence theorem this converges as $\alpha \to 0$ to

$$\sum_{n=1}^{\infty} \mu_n (g, u_n) v_n = K^{\dagger} g$$

(see (1.3.4) and the remark which follows it). #

In case $g \notin \mathcal{D}(K^{\dagger})$ the next result shows that $\{R_\alpha(\tilde{K})K^* g\}_{\alpha>0}$ does not have even a single weakly convergent subsequence.

THEOREM 2.1.2. <u>If</u> $g \notin \mathcal{D}(K^{\dagger})$ <u>then for any sequence</u> $\alpha_n \to 0$, $\{R_{\alpha_n}(\tilde{K})K^* g\}$ <u>is not weakly convergent</u>.

Proof Denote by P the orthogonal projection operator of H_2 onto $\overline{R(K)} = N(K^*)^{\perp} = N(\hat{K})^{\perp}$. If

$$R_{\alpha_n}(\tilde{K})K^* Pg = R_{\alpha_n}(\tilde{K})K^* g \overset{w}{\to} z \in H_1,$$

then by (2),

$$\hat{K} R_{\alpha_n}(\hat{K}) Pg \to Kz$$

since K is compact. However, by (3) and (4) we have

$$\hat{K} R_{\alpha_n}(\hat{K}) Pg \to Pg, \quad \text{as} \quad n \to \infty,$$

i.e., $Pg = Kz$ and hence $g \in \mathcal{D}(K^{\dagger})$. #

These results show that for convergence of the approximations in even a very weak sense it is necessary and sufficient that $g \in \mathcal{D}(K^{\dagger})$. When, in Section 3, we deal with the problem of inaccurate data we will see that this assumption is tantamount to the stronger assumption that $g \in R(K)$ (see also the concluding discussion of the previous chapter).

Since bounded sequences in Hilbert space have weakly convergent subsequences we obtain immediately:

COROLLARY 2.1.3. <u>If</u> $g \notin \mathcal{D}(K^{\dagger})$, <u>then</u> $\lim_{\alpha \to 0} \| R_\alpha(\tilde{K})K^* g \| = \infty$

2.2. Convergence rates

In order to simplify notation we will henceforth denote the approximations studied above by x_α, that is

$$x_\alpha := R_\alpha(\tilde{K})K^*g, \qquad \alpha > 0$$

where R_α is a continuous real-valued function on $[0,\|K\|^2]$ satisfying (3) and (4) of the previous section. The two theorems above illustrate the unequivocal nature of the approximations $\{x_\alpha\}$ in the case where g is known exactly. Before we take up the case of imprecisely known data we will establish some convergence rates for the approximations $\{x_\alpha\}$.

Theorems 2.1.1 and 2.1.2 show that in order to obtain convergence it is necessary and sufficient that $g \in \mathcal{D}(K^\dagger)$, or equivalently that $Pg \in R(K)$, where P is the projection of H_2 onto $\overline{R(K)}$. In order to obtain a rate of convergence it seems natural that we would require a condition somewhat stronger than this. Note that

$$R(K) = R(KP_{N(K)^\perp})$$

where P_S denotes the orthogonal projection onto the closed subspace S. Also, from (1.2.3) it follows that

$$P_{N(K)^\perp}x = P_{N(\tilde{K})^\perp}x = \lim_{\nu\to 0+} \tilde{K}^\nu x$$

and hence it seems not unreasonable to replace the condition $Pg \in R(K)$ (i.e., $g \in \mathcal{D}(K^\dagger)$) with the stronger condition $Pg \in R(K\tilde{K}^\nu)$, some $\nu > 0$, in order to obtain a rate of convergence. To enable us to gauge the rate of convergence we will replace (4) by the condition

$$t^\nu|1-tR_\alpha(t)| \leq \omega(\alpha,\nu) \tag{5}$$

for $t \in [0,\|K\|^2]$, where $\omega(\alpha,\nu)$ is a "rate of convergence" function satisfying

$$\omega(\alpha,\nu) \to 0 \quad \text{as} \quad \alpha \to 0$$

for each $\nu > 0$.

LEMMA 2.2.1. $R(\tilde{K}^\nu) \subseteq N(K)^\perp$.

Proof By (1.2.3) we have

$$\tilde{K}^\nu x = \sum_n \lambda_n^\nu (x, w_n) w_n$$

where λ_n are the nonzero eigenvalues of $\tilde{K}$ and w_n are associated eigenvectors. Therefore

$$w_n = \lambda_n^{-1} \tilde{K} w_n = K^* \lambda_n^{-1} K w_n \in R(K^*),$$

and hence $\tilde{K}^\nu x \in \overline{R(K^*)} = N(K)^\perp$, by (1.2.1). #

THEOREM 2.2.2. <u>If</u> $Pg = K\tilde{K}^\nu w$ <u>for some</u> $\nu > 0$ <u>and some</u> $w \in H_1$, <u>then</u>

$$\| K^\dagger g - x_\alpha \| \leq \omega(\alpha, \nu) \| w \|.$$

Proof Let $u = K^\dagger g$. Then $Ku = Pg = K\tilde{K}^\nu w$ and hence $u - \tilde{K}^\nu w \in N(K)$. However, $u \in N(K)^\perp$ and $\tilde{K}^\nu w \in N(K)^\perp$, therefore $u = \tilde{K}^\nu w$. Also,

$$\begin{aligned} x_\alpha &= R_\alpha(\tilde{K}) K^* g = R_\alpha(\tilde{K}) K^* Pg \\ &= R_\alpha(\tilde{K}) \tilde{K} u = R_\alpha(\tilde{K}) \tilde{K}^{\nu+1} w. \end{aligned}$$

Therefore,

$$\begin{aligned} \| u - x_\alpha \| &= \| \tilde{K}^\nu (I - R_\alpha(\tilde{K})\tilde{K}) w \| \\ &\leq \omega(\alpha, \nu) \| w \| \end{aligned}$$

by (1.2.2) and (1.2.4). #

We point out that, as shown in the proof above, the condition $Pg = K\tilde{K}^\nu w$ implies that $u = K^\dagger g = \tilde{K}^\nu w$. Conversely, if $u = K^\dagger g = \tilde{K}^\nu w$, then by (1.3.1) (iii) we have $Pg = Ku = K\tilde{K}^\nu w$. Therefore the hypothesis $Pg \in R(K\tilde{K}^\nu)$ on the data is equivalent to the hypothesis $u \in R(\tilde{K}^\nu)$ on the solution.

In order to obtain rates of convergence for iterative regularization methods we will find a somewhat stronger hypothesis useful. Namely, we will assume that $Pg \in R(\hat{K}^\nu)$ for some $\nu \geq 1$. Note that this is equivalent to $Pg \in R(K\tilde{K}^{\nu-1}K^*)$ and hence under this condition we obtain immediately from (2.2.2) a rate of convergence of the order $\omega(\alpha, \nu-1)$. However, we now show that a rate of a different sort is obtainable which is strictly better for the cases which interest us. For the sake of a more concise notation, we will denote the error in the approximation x_α by e_α, i.e.

$$e_\alpha := K^\dagger g - x_\alpha .$$

LEMMA 2.2.3. <u>If</u> $Pg = \hat{K}^\nu w$ <u>for some</u> $\nu \geq 1$, <u>then</u> $\|e_\alpha\|^2 \leq \omega(\alpha,\nu-1)\ \|Ke_\alpha\|\,\|w\|$.

Proof As in the previous proof we have $K^\dagger g = K^*\hat{K}^{\nu-1}w$. Also,

$$x_\alpha = R_\alpha(\tilde{K})K^*Pg = K^*R_\alpha(\hat{K})\hat{K}^\nu w.$$

Therefore,

$$e_\alpha = K^*(I - R_\alpha(\hat{K})\hat{K})\hat{K}^{\nu-1}w$$

and hence

$$\begin{aligned}\|e_\alpha\|^2 &= (e_\alpha, K^*(I - R_\alpha(\hat{K})\hat{K})\hat{K}^{\nu-1}w)\\ &= (Ke_\alpha, (I - R_\alpha(\hat{K})\hat{K})\hat{K}^{\nu-1}w)\\ &\leq \omega(\alpha, \nu-1)\|Ke_\alpha\|\ \|w\| . \#\end{aligned}$$

THEOREM 2.2.4. <u>If</u> $Pg = \hat{K}^\nu w$ <u>for some</u> $\nu \geq 1$, <u>then</u>
$\|e_\alpha\| \leq \sqrt{\omega(\alpha,\nu-1)\omega(\alpha,\nu)}\ \|w\|$.

Proof In the proof of (2.2.3) we saw that

$$e_\alpha = K^*(I - R_\alpha(\hat{K})\hat{K})\hat{K}^{\nu-1}w$$

and therefore

$$\tilde{K}e_\alpha = K^*\hat{K}^\nu(I - R_\alpha(\hat{K})\hat{K})w.$$

We then have

$$\begin{aligned}\|Ke_\alpha\|^2 &= (\tilde{K}e_\alpha, e_\alpha) = (\hat{K}^\nu(I - R_\alpha(\hat{K})\hat{K})w, Ke_\alpha)\\ &\leq \omega(\alpha,\nu)\ \|Ke_\alpha\|\ \|w\|,\end{aligned}$$

that is, $\|Ke_\alpha\| \leq \omega(\alpha,\nu)\ \|w\|$.

Lemma 2.2.3 then gives

$$\|e_\alpha\|^2 \leq \omega(\alpha,\nu-1)\omega(\alpha,\nu)\ \|w\|^2 . \#$$

2.3. Regularization with inexact data

In the previous sections convergence results are established for a class of general approximation methods for first kind equations assuming that the data are exactly known. However, it has been repeatedly observed in the previous chapter that the crux of the difficulty in solving first kind equations is that the data are only imprecisely known, that is, only some garbled version g^δ is available satisfying

$$\| g - g^\delta \| \leq \delta$$

where δ is an *a priori* known error level. One is then forced to use the available data to form approximations

$$x_\alpha^\delta := R_\alpha(\tilde{K})K^* g^\delta .$$

These approximations are said to be *regular* if they converge in some sense to the minimal norm solution as $\delta \to 0$. To be more precise, the approximations are regular if there is some choice of the *regularization parameter* α in terms of the noise level δ, say $\alpha = \alpha(\delta)$, such that

$$\lim_{\delta\to 0} R_{\alpha(\delta)}(\tilde{K})K^* g^\delta = K^\dagger g. \qquad (1)$$

That is to say, a *regularization method* consists not only of a choice of regularization functions R_α but also of a choice of norms for the spaces and a choice $\alpha(\delta)$ of the regularization parameter. This last choice may be made in either an *a priori* or an *a postiori* way (see, e.g., [8]) but in any case the mating of the regularization parameter with the noise is at the heart of the theory of regularization.

In order to establish a regularity condition we introduce, in accordance with (3-4) of the previous section, a positive constant C such that

$$|tR_\alpha(t)| \leq C^2, \quad \text{for} \quad t \in [0,\|K\|^2], \quad \alpha > 0 \qquad (2)$$

and a function

$$r(\alpha) = \max\{|R_\alpha(t)| \; : \; t \in [0,\|K\|^2]\}. \qquad (3)$$

Note that, by (3) of the previous section, $r(\alpha) \to \infty$ as $\alpha \to 0$. We now relate the approximation using exact data x_α to the approximation using inexact data x_α^δ.

LEMMA 2.3.1. $\|K(x_\alpha - x_\alpha^\delta)\| \leq \delta C^2$.

Proof Since $\tilde{K}(x_\alpha - x_\alpha^\delta) = \tilde{K}R_\alpha(\tilde{K})K^*(g - g^\delta)$, we have by use of (2),

$$\|K(x_\alpha - x_\alpha^\delta)\|^2 = (\tilde{K}(x_\alpha - x_\alpha^\delta),\ x_\alpha - x_\alpha^\delta)$$

$$= (\tilde{K}R_\alpha(\tilde{K})K^*(g - g^\delta),\ x_\alpha - x_\alpha^\delta)$$

$$= (\hat{K}R_\alpha(\hat{K})(g - g^\delta),\ K(x_\alpha - x_\alpha^\delta))$$

$$\leq \delta C^2 \|K(x_\alpha - x_\alpha^\delta)\|. \quad \#$$

LEMMA 2.3.2. $\|x_\alpha - x_\alpha^\delta\| \leq \delta C\sqrt{r(\alpha)}$.

Proof Since $x_\alpha - x_\alpha^\delta = K^*R_\alpha(\hat{K})(g - g^\delta)$, we have, by use of (3) and (2.3.1),

$$\|x_\alpha - x_\alpha^\delta\|^2 = (x_\alpha - x_\alpha^\delta,\ K^*R_\alpha(\hat{K})(g - g^\delta))$$

$$= (K(x_\alpha - x_\alpha^\delta),\ R_\alpha(\hat{K})(g - g^\delta))$$

$$\leq \delta^2 C^2\ r(\alpha). \quad \#$$

We may now prove a sufficient condition for regularity of the approximations in the presence of error in the data, that is, a condition on $\alpha(\delta)$ so that

$$\lim_{\delta \to 0} x_{\alpha(\delta)}^\delta = K^\dagger g.$$

Here we assume that $\alpha : [0,\infty) \to [0,\infty)$ is a continuous nonnegative parameter choice function with $\alpha(0) = 0$.

THEOREM 2.3.3. _Suppose_ $g \in \mathcal{D}(K^\dagger)$, $\alpha(\delta) \to 0$ _and_ $\delta^2 r(\alpha(\delta)) \to 0$ _as_ $\delta \to 0$, _then_ $x_{\alpha(\delta)}^\delta \to K^\dagger g$ _as_ $\delta \to 0$.

Proof Let $u = K^\dagger g$; we then have, by (2.3.2),

$$\|u - x_{\alpha(\delta)}^\delta\| \leq \|u - x_{\alpha(\delta)}\| + \|x_{\alpha(\delta)} - x_{\alpha(\delta)}^\delta\|$$

$$\leq \|u - x_{\alpha(\delta)}\| + \delta\sqrt{r(\alpha(\delta))}\ C.$$

But $\|u - x_{\alpha(\delta)}\| \to 0$ (by (2.1.1)) and $\delta\sqrt{r(\alpha(\delta))} \to 0$ as $\delta \to 0$. #

We now consider the concept of weak regularity as developed by Engl [4]. Engl's theory will be exposed in a slightly more special and simpler context which is better suited to our purpose. The reader may consult [4] for a more general treatment. The approximations $\{x^{\delta}_{\alpha(\delta)}\}$ are said to be *weakly regular* if for each sequence $\{\delta_n\}$ with $\delta_n \to 0$,

$$x^{\delta_n}_{\alpha(\delta_n)} \stackrel{w}{\to} K^{\dagger}g$$

where $\stackrel{w}{\to}$ indicates weak convergence in the Hilbert space H_1.

In the previous theorem we saw that $\delta^2 r(\alpha(\delta)) \to 0$ was a sufficient condition for strong regularity. We now show that the weaker condition

$$\limsup_{\delta \to 0} \delta^2 r(\alpha(\delta)) < \infty \tag{4}$$

is a sufficient condition for weak regularity.

THEOREM 2.3.4. Suppose that (4) is satisfied and that $g \in \mathcal{D}(K^{\dagger})$. Then the approximations $\{x^{\delta}_{\alpha(\delta)}\}$ are weakly regular.

Proof Again let $u = K^{\dagger}g$ and suppose $\delta_n \to 0$. We then have

$$u - x^{\delta_n}_{\alpha(\delta_n)} = u - x_{\alpha(\delta_n)} + x_{\alpha(\delta_n)} - x^{\delta_n}_{\alpha(\delta_n)}$$

and $x_{\alpha(\delta_n)} \to u$ as $n \to \infty$ by (2.1.1) since by assumption $\alpha(\delta_n) \to 0$. Therefore it is sufficient to show that

$$x_{\alpha(\delta_n)} - x^{\delta_n}_{\alpha(\delta_n)} \stackrel{w}{\to} 0 \text{ as } n \to \infty. \tag{5}$$

By (2.3.2) we have

$$\| x_{\alpha(\delta_n)} - x^{\delta_n}_{\alpha(\delta_n)} \|^2 \leq C^2 \delta_n^2 \, r(\alpha(\delta_n))$$

and hence $\{x_{\alpha(\delta_n)} - x^{\delta_n}_{\alpha(\delta_n)}\}$ is a bounded sequence by (4). To establish (5) it is therefore sufficient (by the Banach-Steinhaus theorem) to show that

$$(z, x_{\alpha(\delta_n)} - x^{\delta_n}_{\alpha(\delta_n)}) \to 0 \quad \text{as} \quad n \to \infty \tag{6}$$

for each z in a dense subset of H_1. Since

$$x_{\alpha(\delta_n)} - x^{\delta_n}_{\alpha(\delta_n)} = R_{\alpha(\delta_n)}(\tilde{K})K^*(g - g^{\delta_n})$$

$$= K^* R_{\alpha(\delta_n)}(\hat{K})(g - g^{\delta_n}),$$

we see that (6) holds trivially for $z \in N(K)$. Also, if $z \in R(K^*)$ say $z = K^*w$, then

$$(z, x_{\alpha(\delta_n)} - x^{\delta_n}_{\alpha(\delta_n)}) = (w, \hat{K}R_{\alpha(\delta_n)}(\hat{K})(g - g^{\delta_n}))$$

$$\leq \|w\| C^2 \delta_n \to 0 \quad \text{as} \quad n \to \infty.$$

Therefore, in view of (1.2.1), (6) holds for all z in the dense subspace

$$R(K^*) + R(K^*)^\perp = R(K^*) + N(K)$$

of H_1, which completes the proof. #

In [4] Engl also introduced the notions of strong and weak divergence. Roughly speaking, the approximations are said to be *strongly divergent* if they may fail to converge even in the *weak* sense and are said to be *weakly divergent* if they may fail to converge in the *strong* sense. Thus in speaking of divergence, the modifiers "strong" and "weak" refer not to the topology in which the divergence may occur but rather to the fact that strong divergence is a more serious condition than weak divergence, i.e., strong divergence implies weak divergence. Therefore in this context it is possible for the approximations to be simultaneously weakly regular and weakly divergent. The precise definitions of strong and weak divergence are given in the conclusions of the two theorems below, respectively.

In the next two theorems we shall suppose that $\alpha : [0,\infty) \to [0,\infty)$ is a continuous strictly increasing function with $\alpha(0) = 0$. We now show that the condition

$$\lim_{\delta\to 0} \delta^2 \alpha(\delta) R_{\alpha(\delta)}(\alpha(\delta))^2 = \infty \tag{7}$$

is sufficient to ensure that the approximations are strongly divergent.

THEOREM 2.3.5. Suppose that K does not have finite rank, $\alpha(\delta)$ satisfies the conditions above and $g \in \mathcal{D}(K^\dagger)$. Then there is a sequence $\{\delta_n\}$ with $\delta_n \to 0$ and vectors g^{δ_n} with $\| g - g^{\delta_n}\| \leq \delta_n$ such that $\{x_{\alpha(\delta_n)}^{\delta_n}\}$ is not weakly convergent.

Proof Recall that by $x_{\alpha(\delta_n)}^{\delta_n}$ we mean the approximation

$$x_{\alpha(\delta_n)}^{\delta_n} = R_{\alpha(\delta_n)}(\tilde{K})K^* g^{\delta_n}.$$

Let $\{u_n, v_n; \mu_n\}$ be a singular system for K and let $\lambda_n = \mu_n^{-2}$. Since K does not have finite rank, we have $\lambda_n \to 0$ as $n \to \infty$. The conditions on α guarantee the existence of a sequence $\{\delta_n\}$ with $\delta_n \to 0$ and $\alpha(\delta_n) = \lambda_n$. Let $g^{\delta_n} = g + \delta_n v_n$. Note that

$$x_{\alpha(\delta_n)} = R_{\alpha(\delta_n)}(\tilde{K})K^* g \to K^\dagger g \quad \text{as} \quad n \to \infty \tag{8}$$

by (2.1.1). Also,

$$x_{\alpha(\delta_n)} - x_{\alpha(\delta_n)}^{\delta_n} = R_{\alpha(\delta_n)}(\tilde{K})K^*(g - g^{\delta_n})$$

$$= \delta_n \mu_n^{-1} R_{\alpha(\delta_n)}(\lambda_n)u_n.$$

Therefore, by (7),

$$\| x_{\alpha(\delta_n)} - x_{\alpha(\delta_n)}^{\delta_n} \|^2 = \delta_n^2 \lambda_n R_{\alpha(\delta_n)}(\lambda_n)^2$$

$$= \delta_n^2 \alpha(\delta_n) R_{\alpha(\delta_n)}(\alpha(\delta_n))^2 \to \infty$$

as $n \to \infty$. By (8) we therefore have

$$\| x_{\alpha(\delta_n)}^{\delta_n} \| \to \infty \quad \text{as} \quad n \to \infty$$

and hence $\{x_{\alpha(\delta_n)}^{\delta_n}\}$ is not weakly convergent. #

Finally, we show that if condition (7) is changed to

$$\liminf_{\delta\to 0} \delta^2\alpha(\delta)R_{\alpha(\delta)}(\alpha(\delta))^2 > 0 \tag{9}$$

then the approximations become weakly divergent.

THEOREM 2.3.6. Suppose that K does not have finite rank, (9) holds and $g \in \mathcal{D}(K^\dagger)$. Then there is a sequence $\{\delta_n\}$ with $\delta_n \to 0$ and vectors g^{δ_n} with $\| g - g^{\delta_n}\| \leq \delta_n$ such that $\{x^{\delta_n}_{\alpha(\delta_n)}\}$ does not converge strongly to $K^\dagger g$.

Proof We proceed with exactly the same construction as in the previous proof. By (9) there is a constant $A > 0$ such that

$$\| x_{\alpha(\delta_n)} - x^{\delta_n}_{\alpha(\delta_n)} \|^2 = \delta_n^2\alpha(\delta_n)R_{\alpha(\delta_n)}(\alpha(\delta_n))^2 \geq A$$

for all n sufficiently large. But

$$x_{\alpha(\delta_n)} \to K^\dagger g \quad \text{as} \quad n \to \infty$$

by (2.1.1) and hence it is impossible for $\{x^{\delta_n}_{\alpha(\delta_n)}\}$ to converge to $K^\dagger g$. #

2.4. Some examples

As representative examples of the preceding theory we will consider regularization methods of the continuous, iterative and expansion types, respectively.

The first method is generated by the function

$$R_\alpha(t) = (t + \alpha)^{-1}, \qquad \alpha > 0,$$

that is,

$$x_\alpha = (\tilde{K} + \alpha I)^{-1}K^*g.$$

This is called *Tikhonov regularization* and all of the subsequent chapters will deal with this method exclusively. For this method we may use the rate of convergence function

$$\omega(\alpha,\nu) = \alpha^\nu \quad \text{for} \quad 0 < \nu \leq 1 \tag{1}$$

(see (5) of Section 2). Also, since

$$|tR_\alpha(t)| \lesssim 1 \quad \text{and} \quad \max|R_\alpha(t)| = \alpha^{-1},$$

we have $C = 1$ and $r(\alpha) = 1/\alpha$ in (2) and (3) of the previous section. Finally, note that

$$\alpha\, R_\alpha(\alpha)^2 = \frac{1}{4\alpha}\,.$$

Making these identifications in the previous section, we may summarize Theorems 2.3.3-6 in the following way:

THEOREM 2.4.1. Let $M = \limsup_{\delta\to 0} \delta^2/\alpha(\delta)$ and $m = \liminf_{\delta\to 0} \delta^2/\alpha(\delta)$. Then $\{x^\delta_{\alpha(\delta)}\}$ is strongly regular, weakly regular, strongly divergent or weakly divergent, respectively, according as $M = m = 0$, $M < \infty$, $M = m = \infty$, or $m > 0$, respectively.

Tikhonov [14] established the weak regularity of the approximations under the assumption that $C_1\alpha \lesssim \delta^2 \lesssim C_2\alpha$ for some positive constants C_1 and C_2. The strong regularity condition above appears in a slightly different but equivalent form in Tikhonov and Arsenin [15].

Our next example of a regularization method is the Landweber-Fridman iteration ([10], [5]) which is defined by

$$x_o = aK^*g, \quad x_{n+1} = (I - a\tilde{K})x_n + aK^*g$$

where a is a constant satisfying $0 < a < 2\|K\|^{-2}$. That is, the method is generated by the function

$$R_n(t) = a \sum_{k=0}^{n} (1 - at)^k.$$

Here the regularization parameter is an iteration number $n = n(\delta)$. Nevertheless this situation may be easily fitted into the framework developed above by setting $\alpha(\delta) = 1/\beta(\delta)$ where $\beta(\delta)$ is a continuous function with $[\beta(\delta)] = n(\delta)$.

For this regularization method one can easily show that we may take

$$C = 1, \qquad r(n) = a(n+1)$$

and with $\alpha(\delta)$ defined as above

$$\alpha(\delta)R_{\alpha(\delta)}(\alpha(\delta))^2 \sim n(\delta).$$

Therefore, setting

$$M = \limsup_{\delta\to 0} \delta^2 n(\delta) \quad \text{and} \quad m = \liminf_{\delta\to 0} \delta^2 n(\delta)$$

we find that the conclusion of Theorem 2.4.1 also holds for this iterative regularization method (see [4, Theorem 4.4]).

As a final example we consider the method of spectral cut-off, or truncated singular function expansion ([1], [2], [12]). In this example one takes

$$x_\alpha = \sum_{n=1}^{[\frac{1}{\alpha}]} \mu_n (g, u_n) v_n$$

where $\{u_n, v_n; \mu_n\}$ is a singular system for K (see (1.3.4)). These approximations are generated by a continuous function $R_\alpha(t)$ which satisfies

$$R_\alpha(t) = \begin{cases} 1/t & , t \geq \mu_{[\frac{1}{\alpha}]}^{-2} \\ 0 & , t < \mu_{[\frac{1}{\alpha}]+1}^{-2} \end{cases} .$$

Indeed, for this function

$$\begin{aligned} R_\alpha(\tilde{K})K^* g &= R_\alpha(\tilde{K})K^* \sum_{n=1}^{\infty} (g, u_n) u_n \\ &= \sum_{n=1}^{\infty} \mu_n^{-1} R_\alpha(\mu_n^{-2})(g, u_n) v_n \\ &= \sum_{n=1}^{[\frac{1}{\alpha}]} \mu_n (g, u_n) v_n . \end{aligned}$$

For this function one has $C = 1$, $r(\alpha) = 1/\alpha$ and $\alpha R_\alpha(\alpha)^2 = 1/\alpha$ and hence Theorem 2.4.1 applies directly to the truncated singular function expansion method.

We point out that the results of this chapter are also applicable to many other types of iterative and non-iterative regularization methods (see, e.g., [4], [13], [3], [6], [11], [9]).

REFERENCES

1. Baker, C.T.H., Fox, L., Mayers, D.F. and Wright, K., Numerical solution of Fredholm integral equations of the first kind, Comput. J. 7(1964), 141-147.
2. Cuppen, J.J.M., Convergence rates for filtered least squares minimal norm approximations of the solution of ill-posed problems, manuscript, Amsterdam, 1980.
3. Dorofeev, I.F., The accuracy of some regularizing algorithms, USSR Comp. Math. Phys. 16(1976), 224-229.
4. Engl, H.W., Necessary and sufficient conditions for convergence of regularization methods for solving linear operator equations of the first kind, Numer. Funct. Anal. & Optimiz. 3(1981), 201-222.
5. Fridman, V., Method of successive approximations for Fredholm integral equations of the first kind, Uspehi Mat. Nauk 11(1956), 233-234 (in Russian).
6. Friedrich, V., Hofmann, B., and Tautenhahn, U., Möglichkeiten der Regularisierung bei der Auswertung von Messdaten, Wissenschaftliche Schriftenreihe der Technischen Hochschule Karl-Marx-Stadt, vol.10, Karl Marx Stadt, 1979.
7. Groetsch, C.W., On a class of regularization methods, Boll. Un. Mat. Ital., Ser. 17-B (1980), 1411-1419.
8. Groetsch, C.W., The parameter choice problem in linear regularization: a mathematical introduction, in "Ill-posed Problems: Theory and Practice," M.Z. Nashed (Ed.), Reidel, Dordrecht, to appear.
9. King, J.T. and Chillingworth, D., Approximation of generalized inverses by iterated regularization, Numer. Func. Anal. & Optimiz. 2(1979), 449-513.
10. Landweber, L., An iteration formula for Fredholm integral equations of the first kind, Amer. J. Math. 73(1951), 615-624.

11. Lardy, L.J., A series representation for the generalized inverse of a closed linear operator, Atti Accad. Naz. Lincei Rend. Cl. Sci. Fis. Mat. Natur., Ser. VIII, 58(1975), 152-157.

12. Lee, J.W. and Prenter, P.M., An analysis of the numerical solution of Fredholm integral equations of the first kind, Numer. Math. 30(1978), 1-23.

13. Strand, O.N., Theory and methods related to the singular function expansion and Landweber's iteration for integral equations of the first kind, SIAM J. Numer. Anal. 11(1974), 798-825.

14. Tikhonov, A.N., Regularization of incorrectly posed problems, Soviet Math. Doklady 4(1963), 1624-1627.

15. Tikhonov, A.N. and Arsenin, V.Y., Solutions of Ill-posed Problems, Wiley, New York, 1977.

3 Tikhonov regularization

In this chapter we delve more deeply into the specific regularization method proposed by Tikhonov and some of its variants. We begin by placing Tikhonov's classical setting within an abstract Hilbert space framework and presenting some basic results. We then look more closely at convergence questions and exhibit some "saturation" and converse results pertaining to Tikhonov regularization. A general technique for choosing the regularization parameter, the discrepancy principle, is studied in Section 3.3. Finally we consider the use of regularizing semi-norms generated by differential operators.

3.1. Tikhonov's method

The main difficulty in solving a Fredholm integral equation of the first kind,

$$Ku(x) := \int_a^b k(x,s)u(s)ds = g(x), \tag{1}$$

arises from the instability of the (generalized) inverse operator. In particular, as the Riemann-Lebesgue lemma shows, certain highly oscillatory noise in a solution may be screened out by the integral operator, giving a result which is very close to the right-hand side g. Tikhonov [30] proposed to damp out such oscillations and "regularize" the solution process by taking as an approximate solution the function z which minimizes

$$\int_a^b (Kz(s) - g(s))^2 ds + \alpha \int_a^b [p(s)z(s)^2 + q(s)z'(s)^2] ds \tag{2}$$

where p and q are strictly positive functions and α is a positive parameter. The duty of this parameter is to effect a trade-off between smoothness (large α) and fidelity (small α) in the approximate solution.

This idea may be phrased abstractly as the problem of minimizing the functional

$$F_\alpha(z) = \|Kz - g\|_2^2 + \alpha \|z\|_1^2 \tag{3}$$

where K is a compact operator from a real Hilbert space H_1 into a real Hilbert space H_2. It is easy to see that the quadratic functional F_α has a unique minimum z_α which is characterized by the condition $f'(0) = 0$ for every vector $w \in H_1$, where

$$f(t) = F_\alpha(z_\alpha + tw).$$

However, an elementary calculation gives

$$f'(0) = 2(K^*Kz_\alpha - K^*g + \alpha z_\alpha, w)$$

where $K^* : H_2 \to H_1$ is the adjoint of K. Therefore the unique minimizer z_α of (3) satisfies

$$K^*Kz_\alpha + \alpha z_\alpha = K^*g. \tag{4}$$

Consider again (2), where for simplicity we take $p = q = 1$. Tikhonov characterized the minimizer z of (2) as the solution of the integro-differential boundary value problem

$$\int_a^b \bar{k}(s,w)z(w)dw - \int_a^b k(w,s)g(w)dw + \alpha(z(s) - z''(s)) = 0, \tag{5}$$

$z'(a) = z'(b) = 0$, where $\bar{k}(s,w)$ is the iterated kernel

$$\bar{k}(s,w) = \int_a^b k(v,s)k(v,w)dv.$$

In order to cast this result in an abstract Hilbert space setting, consider first a compact operator K from a Hilbert space H into a Hilbert space H_2. Let B be a closed, symmetric, positive definite operator defined on a dense subspace $\mathcal{D}(B)$ of H. We may then define an inner product $[\cdot,\cdot]$ on $\mathcal{D}(B)$ by

$$[x,y] = (Bx,y) \tag{6}$$

where $(\cdot,\cdot)$ is the generic inner product on H. Let H_1 be the Hilbert space which is the completion of $\mathcal{D}(B)$ with respect to the inner product (6). Finally, let $K^* : H_2 \to H_1$ be the adjoint of K with respect to the inner product (6) and let $K^{\#} : H_2 \to H$ be the adjoint of K with respect to the inner product $(\cdot,\cdot)$. For any $x \in \mathcal{D}(B)$ and any $y \in H_2$ we then have

$$(x,K^{\#}y) = (Kx,y) = [x,K^*y] = (Bx, K^*y) = (x, BK^*y)$$

and hence $K^{\#} = BK^*$, since $\mathcal{D}(B)$ is dense in H. Since B is positive definite we find that for $z_\alpha \in H_1$, the condition (4) is equivalent to

$$K^{\#}Kz_\alpha + \alpha Bz_\alpha = K^{\#}g \tag{7}$$

Now let $H = H_2 = L^2[a,b]$, $\mathcal{D}(B) = \{\phi : \phi'$ is absolutely continuous, $\phi'(a) = \phi'(b) = 0\}$ and define B by

$$B\phi = \phi - \phi'';$$

then the inner product on $\mathcal{D}(B)$ is given by

$$[\phi,\psi] = (\phi - \phi'', \psi) = (\phi,\psi) + (\phi', \psi')$$

where $(\cdot,\cdot)$ is the L^2-inner product and the corresponding norm is

$$\|\phi\|_1^2 = \|\phi\|^2 + \|\phi'\|^2 ,$$

where $\|\cdot\|$ is the L^2- norm.

The unique minimizer z_α of (3) in H_1 satisfies (7), as shown above. However, if z satisfies (5) then under the interpretation above z satisfies (7), i.e. Tikhonov's solution (5) is given by (7) or equivalently (4). Therefore Tikhonov's context is completely subsumed by the abstract framework above.

We point out that if k is continuous and (1) has a unique solution, then z_α above is in fact the minimizer of (3) over the space $C^1[a,b]$ with no boundary conditions imposed (see Baker [3]) and hence the boundary conditions are in a sense suppressible in that they are not explicitly involved in the minimization of (3).

Higher order Tikhonov regularization may be incorporated into this abstract framework in the same way, by a proper choice of the Hilbert space H_1. That is, the solution of (1) using p^{th} order regularization involves minimizing (3) where

$$\|z\|_1^2 = \sum_{i=0}^{p} \|q_i z^{(i)}\|^2$$

and $q_i \in C^i[a,b]$ are given positive functions. In other words, one works in the Hilbert space H_1 which is the completion of

$$\{z : z^{(i)} \in L^2[a,b],\ i = 0,\ldots,p-2,\ z^{(p-1)} \text{ absolutely continuous}\}$$

under the inner product

$$[x,y] = (Bx,y)$$

where

$$B\phi = \sum_{i=0}^{p} (-1)^i \frac{d^i}{ds^i}\left(q_i(s)\frac{d^i\phi}{ds^i}\right)$$

and as before the minimizer of (3) may be written abstractly and succinctly as the function z_α satisfying (4).

We will henceforth deal with the abstract setting, defining regularized approximations x_α to the minimal norm least squares solution by

$$K^*K x_\alpha + \alpha x_\alpha = K^*g$$

or equivalently

$$x_\alpha = (\tilde{K} + \alpha I)^{-1}K^*g$$

where K^* is the adjoint of the compact linear operator $K : H_1 \rightarrow H_2$ and $\tilde{K} := K^*K$. As we pointed out previously, this fits into the general scheme developed in the previous chapter by setting

$$R_\alpha(t) = (t + \alpha)^{-1}.$$

As we saw in Chapter 2, the function

$$\omega(\alpha,\nu) = \alpha^{-\nu}, \qquad 0 < \nu \leq 1$$

satisfies the requirements of a rate of convergence function. We therefore obtain the following immediate corollaries of (2.2.2) and (2.2.4), respectively.

COROLLARY 3.1.1. <u>If</u> $K^\dagger g \in R(\tilde{K}^\nu)$ <u>for some</u> ν <u>with</u> $0 < \nu \leq 1$, <u>then</u>

$$\| K^\dagger g - x_\alpha \| = O(\alpha^\nu).$$

COROLLARY 3.1.2. <u>If</u> $K^\dagger g \in R(K^*)$, <u>then</u> $\| K^\dagger g - x_\alpha \| = O(\alpha^{1/2})$.

Corollary 3.1.1 for $\nu = 1$ and Corollary 3.1.2 were obtained by Morozov [23] under the assumption that $N(K) = \{0\}$ and $g \in R(K)$.

We see that the fastest rate which is guaranteed by these results is $O(\alpha)$, which occurs when $K^\dagger g \in R(\tilde{K})$. In the next section we will see that $O(\alpha)$ is in fact the fastest rate possible and that if this rate is attained, then it is necessarily the case that $K^{\dagger}g \in R(\tilde{K})$.

We now consider the approximations

$$x_{\alpha}^{\delta} := (\tilde{K} + \alpha I)^{-1}K^{*}g^{\delta}$$

where g^{δ} is an approximation to g satisfying $\| g - g^{\delta}\| \leqslant \delta$. Recalling that $r(\alpha) = \alpha^{-1}$ and $C = 1$, we have from (2.3.2):

$$\| x_{\alpha}^{\delta} - K^{\dagger}g\| \leqslant \| K^{\dagger}g - x_{\alpha}\| + \delta/\sqrt{\alpha}.$$

From this, (3.1.1) and (3.1.2) we obtain:

COROLLARY 3.1.3. <u>If</u> $K^{\dagger}g \in R(K^{*})$ <u>and</u> $\alpha = A\delta$ <u>for some</u> $A > 0$, <u>then</u> $\| K^{\dagger}g - x_{\alpha}^{\delta}\| = O(\sqrt{\delta})$.

COROLLARY 3.1.4. <u>If</u> $K^{\dagger}g \in R(\tilde{K}^{\nu})$ <u>for some</u> ν <u>with</u> $0 < \nu \leqslant 1$ <u>and</u> $\alpha = A\delta^{2/(2\nu+1)}$, <u>then</u> $\| K^{\dagger}g - x_{\alpha}^{\delta}\| = O(\delta^{2\nu/(2\nu+1)})$.

The fastest rate guaranteed by this corollary is $O(\delta^{2/3})$, which occurs when $K^{\dagger}g \in R(\tilde{K})$ and $\alpha = A\delta^{2/3}$. In the next section we will also find that this is the fastest rate possible for nondegenerate kernels and prove a converse result.

Miller [21] has studied the situation where the additional information on the solution takes the form of a prescribed bound rather than a smoothness condition. Consider a compact linear operator $K : H_1 \to H_2$ and suppose $g \in R(K)$. Suppose that the available approximation to g is a vector g^{δ} satisfying $\| g - g^{\delta}\| \leqslant \delta$ and assume that $\| K^{\dagger}g\|_1 \leqslant E$, where E is an *a priori* known bound. Miller's least squares method consists of taking as an approximation the minimizer of the functional

$$F_{\delta}(x) = \| Kx - g^{\delta}\|_2^2 + \left(\frac{\delta}{E}\right)^2 \| x\|_1^2 . \tag{8}$$

We have seen that this minimizer is given by

$$x^{\delta} := (\tilde{K} + \left(\frac{\delta}{E}\right)^2 I)^{-1}K^{*}g^{\delta}. \tag{9}$$

According to (2.4.1) these approximations are weakly regular. This can also be seen directly in the following way (see Bertero [4]): If $\delta_n \to 0$ and

$\| g - g^{\delta_n} \| \leq \delta_n$, then since x^{δ_n} minimizes (8),

$$\left(\frac{\delta}{E}\right)^2 \| x^{\delta_n} \|_1^2 \leq F_{\delta_n}(x^{\delta_n}) \leq F_{\delta_n}(K^\dagger g) \leq 2\delta_n^2 \tag{10}$$

Therefore $\{x^{\delta_n}\}$ is bounded and hence each subsequence contains a subsequence which converges weakly in H_1. If $x^{\delta_k} \overset{w}{\to} z$, then since K is compact, $Kx^{\delta_k} \to Kz$. But by (10)

$$\| Kx^{\delta_k} - g^{\delta_k} \| \leq F_{\delta_k}(x^{\delta_k}) \to 0$$

and hence $Kz = g$. Since each approximation x^δ lies in $R(K^*)$, we also have $z \in \overline{R(K^*)} = N(K)^\perp$ (since $\overline{R(K^*)}$ is weakly closed). Therefore $z = K^\dagger g$. Thus each subsequence of $\{x^{\delta_n}\}$ has a subsequence converging weakly to $K^\dagger g$, i.e. $x^{\delta_n} \overset{w}{\to} K^\dagger g$.

Note that if K is defined on a Hilbert space H and H_1 is compactly embedded in H, then the approximations are strongly convergent in H.

Consider, for instance, the previous example where $H_2 = L^2[a,b]$ and H_1 is the completion of

$$\mathcal{D}(B) = \{\phi : \phi' \text{ is absolutely continuous}, \phi'(a) = \phi'(b) = 0\}$$

with respect to the norm

$$\| \phi \|_1^2 = (B\phi, \phi) = [\phi, \phi]$$

where $B\phi = \phi - \phi''$ and $(\cdot,\cdot)$ is the L^2- inner product. If we define the operator $D : \mathcal{D}(B) \to H_2 \oplus H_2$ by $D\phi = \phi \oplus \phi'$ and let $\| \cdot \|$ be the usual product norm $H_2 \oplus H_2$, then

$$\| D\phi \| = \| \phi \|_1$$

and $B = D^*D$. D is then the "constraint" operator of Miller [20] and (8) becomes

$$F_\alpha(x) = \| Kx - g^\delta \|_2^2 + \left(\frac{\delta}{E}\right)^2 \| Dx \|^2$$

whose minimizer x^δ satisfies

$$K^\# Kx^\delta + \left(\frac{\delta}{E}\right)^2 D^*D\, x^\delta = K^\# g^\delta$$

where $K^{\#}$ is the L^2- adjoint of K, or equivalently (9) where K^* is the H_1 adjoint of K. In this case H_1 is compactly embedded in $L^2[a,b]$ and hence strong L^2-convergence of the approximations is assured.

A word about the choice $\alpha = \delta^2/E^2$ is in order here. Define a "modulus of regularization" by

$$\rho(\epsilon) = \sup\{\|x\| : \|Kx\|_2 \leq \delta, \|x\|_1 \leq 1\}$$

where $\|\cdot\|$ is the norm on the Hilbert space H, $K : H \to H_2$ and $H_1 \subseteq H$ is the Hilbert space generated by the inner product $[\cdot,\cdot]$ above. (Then $\rho(\delta/E)$ is essentially Miller's [20] best stability estimate.) Also define a modulus of convergence by

$$\sigma(\delta,\alpha) = \sup\{\|x^\delta - x\| : \|g^\delta - Kx\|_2 \leq \delta, \|x\|_1 \leq 1\}.$$

Then $\sigma(\delta,\alpha)$ measures in terms of the norm in H the worst possible deviation in the approximation to a "solution" x in the class $\|Kx - g^\delta\|_2 \leq \delta$, $\|x\|_1 \leq 1$. Then it is not hard to see [7] that

$$E\rho(\delta/E) = \sup\{\|x\| : \|Kx\|_2 \leq \delta, \|x\|_1 \leq E\}$$

and

$$E\sigma(\delta/E,\alpha) = \sup\{\|x^\delta - x\| : \|g^\delta - Kx\|_2 \leq \delta, \|x\|_1 \leq 1\}.$$

Franklin [7] has shown that if $\lambda = \alpha E^2/\delta^2$, then

$$E\rho(\Delta/E) \lesssim E\sigma(\delta/E,\alpha) \lesssim E'\rho(\delta'/E')$$

where $\delta' = (1 + \sqrt{1+\lambda})\delta$ and $E' = (1 + \sqrt{1+1/\lambda})E$. This lower bound is independent of α and hence Franklin argues that a good overall strategy, if all that is known about x is that $\|g^\delta - Kx\| \leq \delta$ and $\|x\|_1 \leq E$, is to choose α so that the upper bound is also independent of α. This choice of $\lambda = 1$, i.e., $\alpha = \delta^2/E^2$, then results in the two bounds on the modulus of convergence being equal except for a factor of $1 + \sqrt{2}$.

3.2. Saturation and converse results

In this section we will probe the boundaries of the results on asymptotic orders of convergence for Tikhonov's method which were established in Section 3.1. The results which we present here show that certain rates are best possible and that the attainment of these rates necessitates that the solution satisfies certain "smoothness" conditions. Our presentation is a simplified and specialized version of theory developed in [12-15].

We begin with the error-free case and consider the approximation of the minimal norm least squares solution of the equation

$$Kx = g \tag{1}$$

by Tikhonov regularization.

In approximation theory there are various "saturation" theorems which, roughly stated, say that if a certain known degree of approximation is surpassed, then the function being approximated is in some sense trivial. Our first result, which shows that the best rate given in (3.1.2) cannot be improved except in a trivial instance, may be viewed in the same vein.

THEOREM 3.2.1. *Suppose* $g \in \mathcal{D}(K^\dagger)$ *and* $\|K^\dagger g - x_\alpha\| = o(\alpha)$. *Then* $K^\dagger g = 0$ *and* $x_\alpha = 0$ *for all* α.

Proof Suppose that $x = K^\dagger g \neq 0$ and let $e_\alpha = x_\alpha - x$. Then

$$(\tilde{K} + \alpha I)e_\alpha = K^* g - (\tilde{K} + \alpha I)x = -\alpha x.$$

Therefore

$$\alpha\|x\| \leqslant (\|K\|^2 + \alpha)\|e_\alpha\| = o(\alpha)$$

and hence $x = 0$. If P is the orthogonal projection operator of H_2 onto $\overline{R(K)}$, then

$$0 = Kx = Pg.$$

But then

$$x_\alpha = (\tilde{K} + \alpha I)^{-1}K^* g = (\tilde{K} + \alpha I)^{-1}K^* Pg = 0. \quad \#$$

Our next theorem is the converse of (3.1.2) for the best case $\nu = 1$.

THEOREM 3.2.2. *If* K *is compact*, $g \in \mathcal{D}(K^\dagger)$ *and* $\|K^\dagger g - x_\alpha\| = O(\alpha)$, *then* $K^\dagger g \in R(\tilde{K})$.

Proof Let $\{u_n, v_n; \mu_n\}$ be a singular system for K and again let $x = K^\dagger g$. Then by (1.3.4)

$$x = \sum_{n=1}^{\infty} \mu_n (Pg, u_n) v_n .$$

Also,

$$x_\alpha = (\tilde{K} + \alpha I)^{-1} K^* Pg = \sum_{n=1}^{\infty} \frac{\mu_n}{1 + \alpha\mu_n^2} (Pg, u_n) v_n.$$

Therefore

$$\| x - x_\alpha \|^2 = \alpha^2 \sum_{n=1}^{\infty} (1 + \alpha\mu_n^2)^{-2} \mu_n^6 |(Pg, u_n)|^2.$$

But, $\| x - x_\alpha \|^2 = O(\alpha^2)$ by assumption and hence

$$\sum_{n=1}^{\infty} (1 + \alpha\mu_n^2)^{-2} \mu_n^6 |(Pg, u_n)|^2$$

is bounded as $\alpha \to 0$ and from this it follows that

$$\sum_{n=1}^{\infty} \mu_n^6 |(Pg, u_n)|^2 < \infty.$$

There $Pg \in R(KK^*K)$ by (1.2.7). But this implies that $K^\dagger g \in R(\tilde{K})$. #

The two theorems above show clearly that $O(\alpha)$ is the optimal rate of convergence in the case of error-free data. In (3.1.2) it is also shown that the rate $O(\alpha^{1/2})$ obtains when $K^\dagger g \in R(K^*)$. We now show by example that the converse result for this nonoptimal convergence rate does not hold.

Let K be a compact operator with singular system $\{u_n, v_n; n^{1/2}\}$ and let

$$g = Pg = \sum_{n=1}^{\infty} n^{-3/2} u_n.$$

Then (by (1.2.7)) $Pg \notin R(\hat{K})$ since

$$\sum_{n=1}^{\infty} \mu_n^4 |(Pg, u_n)|^2 = \sum_{n=1}^{\infty} \frac{1}{n} = \infty.$$

However,

$$\| x - x_\alpha \|^2 = \alpha \sum_{n=1}^{\infty} \frac{\alpha}{(1 + \alpha n)^2} < \alpha \int_0^\infty \frac{\alpha}{(1 + \alpha t)^2} dt = \alpha,$$

that is, $\| x - x_\alpha \| = O(\alpha^{1/2})$.

We now consider similar questions when the available data g^δ satisfy

$$\| g - g^\delta \| \leq \delta. \tag{2}$$

We will again denote the Tikhonov approximation formed by using this approximate data by x_α^δ, i.e.,

$$x_\alpha^\delta = (\tilde{K} + \alpha I)^{-1} K^* g^\delta.$$

Of course, since the problem is linear the best rate of convergence that we can hope for is $O(\delta)$. In the case when K has finite rank the choice $\alpha = C\delta$ gives this optimal rate. Indeed, if K has finite rank, then $K^\dagger$ is a bounded operator by (1.3.3). Also,

$$x - x_\alpha^\delta = R_\alpha(\tilde{K})K^*(g - g^\delta) + (K^\dagger - R_\alpha(\tilde{K})K^*)g.$$

However, since K has closed range, $R_\alpha(\tilde{K})K^*$ converges in operator norm to $K^\dagger$ and indeed

$$\| K^\dagger - R_\alpha(\tilde{K})K^* \| = O(\alpha)$$

(see [8]). Therefore the choice $\alpha = C\delta$ gives

$$\| x - x_\alpha^\delta \| = O(\delta).$$

In [13] it is shown (by a variant of the argument given in (3.2.4) below) that if a convergence rate of $o(\delta^{2/3})$ results for all choices of g and g^δ satisfying (2), then K must necessarily have finite rank. A related result of Schock [29] shows that if $u \in R(\hat{K}^\nu)$ for some ν with $0 < \nu \leq 1$, $\alpha(\delta) = C\delta^{2/(2\nu+1)}$ and the error has order $o(\delta^{2\nu/(2\nu+1)})$, then K has finite rank.

We will therefore consider below only operators of infinite rank. According to (3.1.4) the choice $\alpha = C\delta^{2/3}$ gives a rate $O(\delta^{2/3})$. We now show that this is best possible. We shall assume that $\alpha = \alpha(\delta) \to 0$ as $\delta \to 0$ and $g \in R(K)$. Again below we set $x = K^\dagger g$.

LEMMA 3.2.3. _If_ $x \neq 0$, _then_ $\alpha(\delta) = O(\| x - x_{\alpha(\delta)}^\delta \|) + O(\delta)$.

Proof Since

$$(\tilde{K} + \alpha(\delta)I)(x - x_{\alpha(\delta)}^\delta) = \alpha(\delta)x + g - g^\delta$$

we have

$$\alpha(\delta)\|x\| \leq \delta + O(\|x - x^{\delta}_{\alpha(\delta)}\|). \quad \#$$

THEOREM 3.2.4. Suppose that K does not have finite rank and $\|x - x^{\delta}_{\alpha(\delta)}\| = o(\delta^{2/3})$, independently of g^{δ} satisfying $\|g - g^{\delta}\| \leq \delta$. Then $x = 0$.

Proof Let $\{u_n, v_n; \mu_n\}$ be a singular system for K. Since K does not have finite rank, $\mu_n \to \infty$ as $n \to \infty$. Let $\delta_n = \mu_n^{-3}$ and $g^{\delta_n} = g + \delta_n u_n$. For simplicity of notation we replace δ_n by δ and $\alpha(\delta_n)$ by α below. Then

$$\begin{aligned} x^{\delta}_{\alpha} - x &= x_{\alpha} - x + x^{\delta}_{\alpha} - x_{\alpha} \\ &= x_{\alpha} - x + \delta(\tilde{K} + \alpha I)^{-1}K^{*}u_n. \end{aligned}$$

Therefore

$$\|x^{\delta}_{\alpha} - x\|^2 = \|x_{\alpha} - x\|^2 + \frac{2\delta\mu_n}{1 + \alpha\mu_n^2}(x_{\alpha} - x, v_n) + \left(\frac{\delta\mu_n}{1 + \alpha\mu_n^2}\right)^2$$

and hence

$$\delta^{-4/3}\|x^{\delta}_{\alpha} - x\|^2 \geq 2\frac{\delta^{-2/3}}{1 + \alpha\delta^{-2/3}}(x_{\alpha} - x, v_n) + (1 + \alpha\delta^{-2/3})^{-2}.$$

Now if $x \neq 0$, then by (3.2.3), $\alpha\delta^{-2/3} \to 0$ as $\delta \to 0$ and therefore by hypothesis

$$0 \geq 2 \limsup_{\delta \to 0} \frac{\delta^{-2/3}(x_{\alpha} - x, v_n)}{1 + \alpha\delta^{-2/3}} + 1.$$

However, since $\|x^{\delta}_{\alpha(\delta)} - x\| = o(\delta^{2/3})$ for all g^{δ} with $\|g - g^{\delta}\| \leq \delta$, we have in particular for $g = g^{\delta}$, $\|x_{\alpha(\delta)} - x\| = o(\delta^{2/3})$ and hence $0 \geq 1$, a contradiction.

We now present a converse to (3.1.4) for the case $\nu = 1$, i.e., the best asymptotic rate.

THEOREM 3.2.5. Suppose $\alpha(\delta) = C\delta^{2/3}$, $C \neq 0$. If $\|x - x^{\delta}_{\alpha(\delta)}\| = O(\delta^{2/3})$ independently of g^{δ} satisfying (2), then $x \in R(\tilde{K})$.

Proof Again let $\{u_n, v_n; \mu_n\}$ be a singular system for K and suppose $g^\delta = (1+\delta)g$ (clearly we may assume that $\|g\| = 1$). Then by (1.3.4)

$$x - x_\alpha^\delta = \sum_{n=1}^{\infty} \{\mu_n(g, u_n) - \frac{\mu_n}{1 + \alpha\mu_n^2} (g^\delta, u_n)\}v_n$$

and hence

$$\| x - x_\alpha^\delta\|^2 = \sum_{n=1}^{\infty} \left\{\frac{\alpha - \delta\mu_n^{-2}}{1 + \alpha\mu_n^2}\right\}^2 \mu_n^6 |(g, u_n)|^2.$$

Therefore there is a constant M such that

$$M \geq \| x - x_\alpha^\delta\|^2 \delta^{-4/3} = \sum_{n=1}^{\infty} \left\{\frac{C - \delta^{1/3}\mu_n^{-2}}{1 + \alpha\mu_n^2}\right\}^2 \mu_n^6 |(g, u_n)|^2$$

$$\geq \sum_{n=1}^{N} \left\{\frac{C - \delta^{1/3}\mu_n^{-2}}{1 + \alpha\mu_n^2}\right\}^2 \mu_n^6 |(g, u_n)|^2$$

for all N. Letting $\delta \to 0$ we find that

$$\sum_{n=1}^{N} C^2\mu_n^6|(g, u_n)|^2 \leq M$$

for all N and hence by (1.2.7) $g \in R(K\tilde{K})$, i.e., $K^\dagger g \in R(\tilde{K})$. #

We note that the corresponding converse of (3.1.3) does not hold. Indeed, for the same example considered previously we have

$$\| x - x_{\alpha(\delta)}^\delta \| \leq \| x - x_\alpha\| + \| x_\alpha - x_{\alpha(\delta)}^\delta\|$$

$$\leq \sqrt{\alpha} + C\delta/\sqrt{\alpha}$$

by (2.3.2). Therefore a choice $\alpha = A\delta$ gives a rate $O(\sqrt{\delta})$, while $K^\dagger g \notin R(K^*)$.

The above results show that the convergence rate $O(\delta^{2/3})$ is the best that can be expected using Tikhonov regularization on a nondegenerate kernel, irrespective of the smoothness of the solution. However, assuming that the solution is sufficiently smooth, even if the kernel is nondegenerate, one can obtain, by use of an iterative method, a rate of convergence which is arbitrarily near to the optimal rate $O(\delta)$ obtainable for degenerate kernels. Consider for example the Landweber-Fridman iteration (see Section 2.4):

$$x_o^\delta = aK^*g^\delta, \quad x_{n+1}^\delta = (I - a\tilde{K})x_n^\delta + aK^*g^\delta$$

where $0 < a < 2\|K\|^{-2}$. For this method one can show that a rate of convergence function is given by

$$\omega(n,\nu) = (n+1)^{-\nu}, \qquad \nu \geq 1$$

and that $C = 1$, $r(n) = a(n+1)$. Therefore, if $g \in R(\hat{K}^\nu)$ for some positive integer ν, then by (2.2.4) and (2.3.2)

$$\|K^\dagger g - x_n^\delta\| \leq C_1(n+1)^{(1-2\nu)/2} + C_2\delta\sqrt{n+1}$$

and hence a choice of iteration number of the form

$$n+1 = [\delta^{-1/\nu}]$$

gives a convergence rate of $O(\delta^{1-1/2\nu})$ which is arbitrarily near $O(\delta)$ for ν sufficiently large.

3.3. The discrepancy principle

We now treat a general *a postiori* strategy for choosing the regularization parameter as a function of the error level which is much favored by Soviet mathematicians. This principle was first clearly enunciated by Morozov, although it had been used earlier in an intuitive way by others. It is based on the reasonable view that the quality of the results of a computation must be comparable to the quality of the input data. To quote from Morozov [23] in translation, "the magnitude of the error must be in agreement with the accuracy of the assignment of the input data."

Suppose then that K is a compact operator, $g \in R(K)$ and we wish to solve

$$Kx = g \tag{1}$$

but we have at our disposal only an approximate right-hand side g^δ which satisfies

$$\|g - g^\delta\| \leq \delta < \|g^\delta\|. \tag{2}$$

Note that this condition is not a genuine restriction. Indeed, the second part of (2) may be simply interpreted as saying that the signal-to-noise ratio, $\|g^\delta\|/\delta$, is strictly greater than one. It is easy to defend the position that if such were not the case, then the data are so hopelessly corrupted as to make any mathematical analysis at all ill-advised.

Morozov's discrepancy principle [22] (see also Ivanov [18]) calls for choosing the regularization parameter $\alpha = \alpha(\delta)$ by the criterion

$$\| Kx_\alpha^\delta - g^\delta \| = \delta \tag{3}$$

in harmony with the maxim that the discrepancy should be in agreement with the error in the input data. The left-hand side of (3) we will call the *discrepancy* in the approximate solution x_α^δ and we will denote it by $D(\alpha;g^\delta)$, i.e.,

$$D(\alpha;g^\delta) := \| Kx_\alpha^\delta - g^\delta \| .$$

We will show below that there is a unique value $\alpha = \alpha(\delta)$ of the regularization parameter satisfying $D(\alpha(\delta); g^\delta) = \delta$. But first we provide some more solid motivation for the choice (3).

If all that is known about the data is (2) and we seek a minimum norm solution of (1), then it would seem reasonable to choose as an approximate solution a vector u in the set

$$C_\delta = \{z : \| Kz - g^\delta \| \leq \delta\}$$

which has minimal norm. Note that since C_δ is closed and convex, it contains a unique vector, say z_δ, of minimum norm. We claim that z satisfies the discrepancy criterion, i.e.,

$$\| Kz_\delta - g^\delta \| = \delta$$

Indeed if this were not the case then we would have

$$\| Kz_\delta - g^\delta \| < \delta$$

and hence for $0 < t < 1$ and sufficiently near 1 we would have

$$\| Ktz_\delta - g^\delta \| < \delta.$$

However, $\| tz_\delta \| < \| z_\delta \|$, contradicting the minimality of z_δ. Therefore it would seem reasonable to choose the parameter $\alpha = \alpha(\delta)$ so that (3) is satisfied.

Before giving further motivation, we establish some basic facts about the discrepancy function $D(\alpha; g^\delta)$.

THEOREM 3.3.1. <u>Suppose that</u> g <u>and</u> g^δ <u>satisfy (2). Then the function</u> $\alpha \to D(\alpha; g^\delta)$ <u>is continuous, increasing and contains</u> δ <u>in its range</u>.

Proof Let $\{u_n, v_n; \mu_n\}$ be a singular system for K. Then since

$$x_\alpha^\delta = (\tilde{K} + \alpha I)^{-1} K^* g^\delta$$

we have

$$Kx_\alpha^\delta - g^\delta = \hat{K}(\hat{K} + \alpha I)^{-1} g^\delta - g^\delta$$

$$= \sum_{n=1}^{\infty} \frac{-\alpha\mu_n^2}{1 + \alpha\mu_n^2} (g^\delta, u_n)u_n - Pg^\delta$$

where P is the orthogonal projection of H_2 onto $N(K^*) = R(K)^\perp$. Therefore

$$D(\alpha; g^\delta)^2 = \sum_{n=1}^{\infty} \left(\frac{\alpha\mu_n^2}{1 + \alpha\mu_n^2} \right)^2 |(g^\delta, u_n)|^2 + \| Pg^\delta \|^2$$

from which it follows that $\alpha \to D(\alpha; g^\delta)$ is increasing and continuous. Note that

$$\lim_{\alpha\to\infty} D(\alpha; g^\delta) = \| (I-P)g^\delta \|^2 + \| Pg^\delta \|^2 = \| g^\delta \|^2 > \delta^2.$$

Also, since $g \in R(K)$ and P projects onto the orthogonal complement of R(K),

$$\lim_{\alpha\to 0} D(\alpha; g^\delta) = \| Pg^\delta \| < \| g - g^\delta \| \leq \delta$$

and hence δ is in the range of $D(\cdot; g^\delta)$. #

From this theorem we conclude that there is a unique value of α, say $\alpha = \alpha(\delta)$, such that

$$D(\alpha(\delta); g^\delta) = \delta. \qquad (4)$$

The choice of α according to this condition is called the choice by the *discrepancy method*.

We can gain more insight into the choice of the regularization parameter by this method in the following way. Let

$$r(\alpha; g^\delta) := g^\delta - Kx_\alpha^\delta$$

represent the residual of the approximation x_α^δ. Note that

$$D(\alpha; g^\delta) = \| r(\alpha; g^\delta) \| \tag{5}$$

and that

$$K^* r(\alpha; g^\delta) = K^* g^\delta - \tilde{K}(\tilde{K} + \alpha I)^{-1} K^* g^\delta$$

$$= \alpha(\tilde{K} + \alpha I)^{-1} K^* g^\delta = \alpha x_\alpha^\delta. \tag{6}$$

Something which is naturally to be desired in the approximations x_α^δ is that they have small error. But note that by (6), (5) and (2) the square error satisfies

$$\| x - x_\alpha^\delta \|^2 = \| x_\alpha^\delta \|^2 - \frac{2}{\alpha} (K^* r(\alpha; g^\delta), x) + \| x \|^2$$

$$= \| x_\alpha^\delta \|^2 - \frac{2}{\alpha} (r(\alpha; g^\delta), g) + \| x \|^2$$

$$= \| x_\alpha^\delta \|^2 - \frac{2}{\alpha} (r(\alpha; g^\delta), g^\delta) + \frac{2}{\alpha} (r(\alpha; g^\delta), g^\delta - g) + \| x \|^2$$

$$\leq E(\alpha; g^\delta)$$

where

$$E(\alpha; g^\delta) := \| x_\alpha^\delta \|^2 - \frac{2}{\alpha} (r(\alpha; g^\delta), g^\delta) + \frac{2\delta}{\alpha} D(\alpha; g^\delta) + \| x \|^2 .$$

Therefore the quantity $E(\alpha; g^\delta)$ is an overestimate of the square error (see Bulter et al. [5] for a similar discussion for constrained least squares problems). Note, however, that $E(\alpha; g^\delta)$ is not always a strict overestimate of the square error. In fact, if g is a norm one eigenvector of $\hat{K}$, associated say with an eigenvalue $\lambda > 0$, and $g^\delta = (1+\delta)g$, then

$$\| x - x_\alpha^\delta \|^2 = E(\alpha; g^\delta)$$

since in this case

$$r(\alpha; g^\delta) = -\frac{\alpha(1+\delta)}{\lambda+\alpha} g$$

and $g^\delta - g = \delta g$ are linearly dependent.

The next theorem gives more feeling for the discrepancy principle by showing that this over-estimate for the square error is minimized precisely when the parameter is chosen according to the discrepancy method (4).

THEOREM 3.3.2. <u>If</u> g <u>and</u> g^δ <u>satisfy (2), then</u> $E(\alpha; g^\delta)$ <u>is a minimum if and only if</u> $D(\alpha; g^\delta) = \delta$.

Proof Note that $D(\alpha; g^\delta)$ is positive for all positive α, for otherwise we have

$$r(\alpha; g^\delta) = Kx_\alpha^\delta - g^\delta = 0$$

for some $\alpha > 0$. But then by (5) we have

$$x_\alpha^\delta = (\tilde{K} + \alpha I)^{-1}K^* g^\delta = 0$$

and hence $g^\delta \in N(K^*) = R(K)^\perp$. Since $g \in R(K)$, we have by (2)

$$\delta^2 \geqslant \| g - g^\delta \|^2 = \| g \|^2 + \| g^\delta \|^2 > \| g \|^2 + \delta^2,$$

a contradiction.

A routine, albeit tedious, calculation gives

$$\frac{d}{d\alpha} E(\alpha; g^\delta) = 2(1-\delta/D(\alpha; g^\delta)) \| (\hat{K} + \alpha I)^{-1/2} r(\alpha; g^\delta) \|^2/\alpha .$$

The factor

$$\| (\hat{K} + \alpha I)^{-1/2} r(\alpha; g^\delta) \|^2/\alpha$$

is always positive for the same reason as above. Since $\alpha \to D(\alpha; g^\delta)$ increases, it follows that

$$\frac{d}{d\alpha} E(\alpha; g^\delta) < 0 \quad \text{for} \quad D(\alpha; g^\delta) < \delta$$

and

$$\frac{d}{d\alpha} E(\alpha; g^\delta) > 0 \quad \text{for} \quad \delta < D(\alpha; g^\delta).$$

Therefore $E(\alpha; g^\delta)$ is a minimum if and only if $\delta = D(\alpha; g^\delta)$. #

We now establish the regularity of the discrepancy method.

THEOREM 3.3.3. <u>If</u> g <u>and</u> g^δ <u>satisfy (2) and</u> $\alpha = \alpha(\delta)$ <u>satisfies (4), then</u> $x_{\alpha(\delta)}^\delta \to x$ <u>as</u> $\delta \to 0$.

Proof Since $x_{\alpha(\delta)}^\delta$ minimizes $F_{\alpha(\delta)}(\cdot; g^\delta)$ we have

$$\| r(\alpha(\delta);g^{\delta})\|^2 + \alpha\| x^{\delta}_{\alpha(\delta)}\|^2 = F_{\alpha(\delta)}(x^{\delta}_{\alpha(\delta)}) \leqslant F_{\alpha(\delta)}(x)$$

$$= \| g - g^{\delta}\|^2 + \alpha\|x\|^2$$

$$\leqslant \delta^2 + \alpha\|x\|^2 .$$

But by (4), $\| r(\alpha(\delta);\ g^{\delta})\| = \delta$, hence

$$\| x^{\delta}_{\alpha(\delta)}\| \leqslant \|x\| \tag{7}$$

for all $\delta > 0$. Therefore for each sequence $\{\delta_n\}$ with $\delta_n \to 0$, there is a subsequence, which we continue to denote by $\{\delta_n\}$, such that

$$x^{\delta_n}_{\alpha(\delta_n)} \overset{w}{\to} y$$

for some $y \in H_1$. Since K is compact,

$$Kx^{\delta_n}_{\alpha(\delta_n)} \to Ky.$$

However, by (4)

$$\| Kx^{\delta_n}_{\alpha(\delta_n)} - g^{\delta_n}\| = \delta_n \to 0$$

and by (2) $g^{\delta_n} \to g$. Therefore $Ky = g$. Since $x^{\delta_n}_{\alpha(\delta_n)} \in R(K^*)$, we have $y \in \overline{R(K^*)} = N(K)^{\perp}$. There y is the minimal norm solution, i.e., $y = x$.

Thus each subsequence of $\{x^{\delta_n}_{\alpha(\delta_n)}\}$ contains a subsequence converging weakly to x and hence

$$x^{\delta_n}_{\alpha(\delta_n)} \overset{w}{\to} x \quad \text{as} \quad n \to \infty.$$

Also, since the norm function is weakly lower semi-continuous, we have by (7)

$$\|x\| \leqslant \liminf_{n\to\infty} \| x^{\delta_n}_{\alpha(\delta_n)}\| \leqslant \limsup_{n\to\infty} \| x^{\delta_n}_{\alpha(\delta_n)}\| \leqslant \|x\|$$

and therefore

$$\| x^{\delta_n}_{\alpha(\delta_n)} \| \to \| x \| .$$

But in Hilbert space weak convergence along with convergence of norms implies strong convergence, and hence

$$x^{\delta}_{\alpha(\delta)} \to x \quad \text{as} \quad \delta \to 0. \quad \#$$

Vinokurov [34] has given a useful *a priori* upper bound for the regularization parameter when chosen by the discrepancy method. In the next theorem we give a very simple proof of Vinokurov's bound.

THEOREM 3.3.4. <u>If</u> $\alpha(\delta)$ <u>is given by (4), then</u> $\alpha(\delta) \leq \delta \| K \|^2 / (\| g^\delta \| - \delta)$.

Proof By (4), (5) and (6) we have

$$\| g^\delta \| - \delta = \| g^\delta \| - \| r(\alpha(\delta); g^\delta) \| \leq \| K x^\delta_{\alpha(\delta)} \|$$

$$\leq \| K \|^2 \| r(\alpha(\delta); g^\delta) \| / \alpha(\delta) = \| K \|^2 \delta / \alpha(\delta). \quad \#$$

We now present a convergence rate, assuming that the minimal norm solution is in $R(K^*)$.

THEOREM 3.3.5. <u>If</u> $x \in R(K^*)$, <u>then</u> $\| x - x^\delta_{\alpha(\delta)} \| = O(\sqrt{\delta})$.

Proof If $x = K^* w$, then by (7), (4) and (2) we have

$$\begin{aligned} \| x^\delta_{\alpha(\delta)} - x \|^2 &= \| x^\delta_{\alpha(\delta)} \|^2 - 2(x^\delta_{\alpha(\delta)}, x) + \| x \|^2 \\ &\leq 2\{ \| x \|^2 - (x^\delta_{\alpha(\delta)}, x) \} \\ &= 2(x - x^\delta_{\alpha(\delta)}, K^* w) \\ &= 2(Kx - Kx^\delta_{\alpha(\delta)}, w) \\ &= 2(g - g^\delta, w) + 2(g^\delta - Kx^\delta_{\alpha(\delta)}, w) \\ &\leq 4\delta \| w \| . \quad \# \end{aligned}$$

The bound obtained above can be improved to $O(\delta)$ if K is an operator of finite rank. In fact, in this case the operator $K^\dagger$ is bounded by (1.3.3). Also, since x and x^δ_α are orthogonal to the nullspace of K we have

$$x^{\delta}_{\alpha(\delta)} - x = K^{\dagger}Kx^{\delta}_{\alpha(\delta)} - K^{\dagger}Kx$$

$$= K^{\dagger}(Kx^{\delta}_{\alpha(\delta)} - g)$$

and therefore

$$\| x^{\delta}_{\alpha(\delta)} - x\| \leqslant \| K^{\dagger}\| \; \{\| Kx^{\delta}_{\alpha(\delta)} - g^{\delta}\| + \| g^{\delta} - g\| \}$$

$$\leqslant 2\| K^{\dagger}\| \, \delta.$$

We now show that for operators with infinite rank, the order of convergence given in (3.3.5) cannot be improved (see [11]).

THEOREM 3.3.6. _If_ $\| x^{\delta}_{\alpha(\delta)} - x\| = o(\sqrt{\delta})$ _for all_ g _and_ g^{δ} _satisfying (2), then_ K _has finite rank._

Proof Let $\{u_n, v_n; \mu_n\}$ be a singular system for K. If K does not have finite rank, then $\mu_n \to \infty$. Let $g = u_1$ and $g^{\delta_n} = u_1 + \delta_n u_n$. Then

$$x = \mu_1 v_1 \quad \text{and} \quad \| g - g^{\delta_n}\| = \delta_n < \| g^{\delta_n}\| .$$

Also,

$$x^{\delta_n}_{\alpha(\delta_n)} - x = (\tilde{K} + \alpha(\delta_n)I)^{-1}K^{*}u_1 - \mu_1 v_1 + \delta_n(\tilde{K} + \alpha(\delta_n)I)^{-1}K^{*}u_n$$

$$= \frac{-\alpha(\delta_n)\mu_1^3}{1 + \alpha(\delta_n)\mu_1^2} v_1 + \frac{\delta_n \mu_n}{1 + \alpha(\delta_n)\mu_n^2} v_n .$$

Therefore, if $\delta_n = \mu_n^{-2}$,

$$\| x^{\delta_n}_{\alpha(\delta_n)} - x\|^2 \geqslant \left(\frac{\delta_n \mu_n}{1 + \alpha(\delta_n)\mu_n^2}\right)^2 = \left(\frac{\sqrt{\delta_n}}{1 + \alpha(\delta_n)/\delta_n}\right)^2$$

and hence, by hypothesis

$$\frac{\sqrt{\delta_n}}{1 + \alpha(\delta_n)/\delta_n} = o(\sqrt{\delta_n}).$$

This says that $\alpha(\delta_n)/\delta_n \to \infty$, but by (3.3.4)

$$\alpha(\delta_n)/\delta_n \leq \|K\|^2/(\sqrt{1+\delta_n^2} - \delta_n) \to \|K\|^2 ,$$

which is a contradiction. #

Independently of and contemporaneously with Morozov, Arcangeli [2] published a discrepancy-like method. Arcangeli proposed to choose the parameter by the condition

$$\| Kx_\alpha^\delta - g^\delta \| = \delta/\sqrt{\alpha} . \tag{8}$$

Note that as in (3.3.1), for fixed $\delta > 0$ the function

$$\phi(\alpha) = \sqrt{\alpha}\ \| Kx_\alpha^\delta - g^\delta \|$$

is continuous, increasing and

$$\lim_{\alpha\to 0} \phi(\alpha) = 0, \qquad \lim_{\alpha\to\infty} \phi(\alpha) = \infty.$$

Therefore there is a unique value $\alpha = \alpha(\delta)$ satisfying (8).

LEMMA 3.3.7. $\alpha(\delta) \to 0$ as $\delta \to 0$.

Proof Suppose $\{\delta_n\}$ is a sequence with $\delta_n \to 0$ and $\alpha(\delta_n) \to C > 0$. Then by (8)

$$0 = \lim_{n\to\infty} \sqrt{\alpha(\delta_n)}\ \| Kx_{\alpha(\delta_n)}^{\delta_n} - g^{\delta_n} \|$$

$$= \sqrt{C}\ \| K(\tilde{K} + CI)^{-1}K^*g - g \| .$$

Therefore

$$(\hat{K} + CI)^{-1}\hat{K}g = g$$

i.e. $\hat{K}g = \hat{K}g + Cg$

and hence $g = 0$, violating (2). #

We now present Arcangeli's regularity theorem.

THEOREM 3.3.8. _If_ $\alpha = \alpha(\delta)$ _is chosen according to (8), then_ $x_{\alpha(\delta)}^\delta \to x$ _as_ $\delta \to 0$.

Proof By (8) we have

$$\delta^2/\alpha(\delta) = \| Kx^\delta_{\alpha(\delta)} - g^\delta \|^2 \leqslant F_{\alpha(\delta)}(x^\delta_{\alpha(\delta)})$$

$$= F_{\alpha(\delta)}(x) = \| Kx - g^\delta \|^2 + \alpha(\delta) \| x \|^2$$

$$\leqslant \delta^2 + \alpha(\delta) \| x \|^2 .$$

Therefore by (3.3.7) $\delta^2/\alpha(\delta) \to 0$ as $\delta \to 0$ and hence

$$x^\delta_{\alpha(\delta)} \to x$$

by (2.4.1). #

3.4. Use of differential operators

In a paper which predates that of Tikhonov, Phillips [26] proposed a Tikhonov-type regularization method for solving a Fredholm equation of the first kind

$$\int_0^1 k(s,t)z(t)dt = g(s) \tag{1}$$

(we take $[a,b] = [0,1]$ for convenience). Phillips' idea was to damp out the highly oscillatory noise in the approximate solution by adding a penalty term of the form $\alpha\| z'' \|^2$ to the usual least squares functional. That is, Phillips takes as an approximate solution the minimizer of the functional

$$F_\alpha(z) := \| Kz - g\|^2 + \alpha\| z'' \|^2 \tag{2}$$

where each norm is the L^2-norm. Note that this differs from Tikhonov regularization in that the regularization term is a seminorm rather than a norm. Also, the Tikhonov functional, unlike (2), always contains a term which tends to minimize the mean of the approximate solution, which may not be a desirable feature.

Note that (2) may not have a unique minimum if N(K) contains a nonzero linear function. Phillips minimized a discrete version of (2) in which he effectively removed the nonuniqueness by imposing zero boundary conditions on the approximate solution.

Before taking up the general case, let us consider the simplest example of regularization using a differentiable operator, i.e., minimization of the functional

$$f_\alpha(z) = \| Kz - g\|^2 + \alpha\| z'\|^2 .$$

Let W be the space of absolutely continuous functions on $[0,1]$. Assuming that N(K) contains no nonzero constant, i.e., that

$$\int_0^1 k(s,t)dt \neq 0,$$

we can show that this functional has a unique minimizer in W in the following way. Let $\mathcal{H}$ represent the Hilbert space $L^2[0,1] \oplus L^2[0,1]$ with the norm

$$|x \oplus y|^2 = \| x \|^2 + \alpha\| y \|^2$$

and associated inner product. Define the injective linear operator $\mathcal{F} : W \to \mathcal{H}$ by $z = Kz + z'$. Then minimizing f_α over W is equivalent to finding a vector of minimum $|\cdot|$-norm in $\mathcal{F}W$. Of course, a unique such vector exists if $\mathcal{F}W$ is closed in $\mathcal{H}$ and this in turn gives a unique minimizer of f_α in W since $\mathcal{F}$ is injective.

However, $\mathcal{F}W$ is indeed closed in $\mathcal{H}$. For if $\{\mathcal{F}u_n\}$ is convergent in $\mathcal{H}$, where $\{u_n\} \subseteq W$, then since each u_n may be uniquely written as $u_n = \lambda_n + \bar{u}_n$, where λ_n is a constant and $\bar{u}_n$ is orthogonal to every constant, i.e.,

$$\int_0^1 \bar{u}_n(t)dt = 0, \tag{3}$$

we find (considering the norm $|\cdot|$ on $\mathcal{H}$) that $\{Ku_n\}$ and $\{\bar{u}_n'\}$ are Cauchy sequences in $L^2[0,1]$. From (3) follows the well-known inequality ([16, p.185])

$$\| \bar{u}_n' \|^2 \geqslant \pi^2 \| \bar{u}_n \|^2$$

and hence $\{\bar{u}_n\}$ is a Cauchy sequence in $L^2[0,1]$. However, since differentiation is a closed operator in W, there is a $z \in W$ with $\bar{u}_n \to z$ and $\bar{u}_n' \to z'$ in $L^2[0,1]$. Now, $Ku_n = \lambda_n K1 + K\bar{u}_n$ and both $\{Ku_n\}$ and $\{K\bar{u}_n\}$ are convergent in $L^2[0,1]$ and hence $\lambda_n \to \lambda$ for some constant λ. Therefore

$$Ku_n \to K(\lambda+z)$$

and

$$u_n' = \overline{u}_n' \to z' = (\lambda + z)'$$

in $L^2[0,1]$, i.e.,

$$\mathcal{F}u_n \to \mathcal{F}(\lambda + z)$$

in $\mathcal{H}$, and hence $\mathcal{F}W$ is closed in $\mathcal{H}$.

Suppose that z_α is the unique minimizer of f_α in W. Then z_α is the unique member of W satisfying

$$\frac{d}{dt} f_\alpha(z_\alpha + tv)\bigg|_{t=0} = 0$$

for all $v \in W$. A direct computation then shows that z_α is the unique member of W satisfying

$$(K^*Kz_\alpha - K^*g, v) + \alpha\int_0^1 z_\alpha'(t)v'(t)dt = 0 \tag{4}$$

for all $v \in W$. Suppose now that $z \in W$ satisfies $z'' \in L^2[0,1]$ and $z'(0) = z'(1) = 0$. Then $z_\alpha = z$ will satisfy (4) if and only if

$$(K^*Kz - K^*g, v) - \alpha(z'', v) = 0$$

for all $v \in W$, i.e., if and only if

$$K^*Kz - \alpha z'' = K^*g. \tag{5}$$

We may now show that a unique such z exists. Note that (5) can have at most one solution satisfying $z'(0) = z'(1) = 0$, for if

$$K^*Kz - \alpha z'' = 0$$

then

$$0 \leq (Kz, Kz) = \alpha(z'', z) = -\alpha(z', z')$$

but $\alpha > 0$ and N(K) contains no nonzero constant by assumption, therefore $z = 0$. To see that a solution actually exists, let $\gamma(s,t)$ be Green's function for the problem $z - z'' = 0$, $z'(0) = z'(1) = 0$, i.e.,

$$\gamma(s,t) = \begin{cases} \cosh(s)\cosh(t-1), & 0 \leq s \leq t \\ \cosh(s-1)\cosh(t), & t \leq s \leq 1 \end{cases}$$

and let G be the integral operator on $L^2[0,1]$ generated by the kernel γ. Then rewriting (5) as

$$(K^*K - \alpha I)z + \alpha(z - z'') = K^*g$$

we see that it is equivalent to

$$G(K^*K - \alpha I)z + \alpha z = GK^*g.$$

Since this is a Fredholm equation of the second kind which has at most one solution, the existence of a solution is guaranteed by the Fredholm alternative theorem ([28, p.170]).

Note that in the minimization of f_α no boundary conditions are explicitly involved, nevertheless the minimizing function satisfies certain boundary conditions. In this sense we may consider the boundary conditions "natural".

Elements of the argument above will reappear in a more general context in the proofs below. We will now consider a more general situation.

Several authors have investigated regularization with differential operators ([17], [19], [20], [27], [32]). Our presentation follows the developments of Lukas [20] and Locker-Prenter [19].

Define a subspace W of $L^2[0,1]$ as follows:

$$W = \{u: u^{(i)} \in C[0,1],\ i = 0,\ldots,m-2,\ u^{(m-1)} \text{ absolutely continuous}\}.$$

Let $w_i \in C^i[0,1]$, $i = 0,\ldots,m-1$, $1/w_m \in C[0,1]$ be given weight functions and define an mth order differential operator $D : W \to L^2[0,1]$ by

$$Du = \sum_{i=0}^{m} w_i u^{(i)}.$$

Equation (1) will be regularized by minimizing the functional

$$F_\alpha(z) = \|Kz - g\|^2 + \alpha\|Dz\|^2 \tag{6}$$

over W where $\|\cdot\|$ indicates the L^2-norm. Note that since $\|D\cdot\|$ is merely a seminorm, this is quite different from Tikhonov regularization. Also note that, as an operator on $L^2[0,1]$, D is closed, densely defined and has a finite dimensional nullspace and closed range.

The argument given above for the simple case may be extended to prove the existence of a minimizer of (4) (see [27], [20]). However, this approach gives no useful characterization of the minimizer. We now present a result which gives existence as well as a natural characterization of the minimizer.

THEOREM 3.4.1. The functional (4) has a minimizer in W which is unique if $N(K) \cap N(D) = \{0\}$. In this case the minimizer is the unique element x_α of $\mathcal{D}(D^*D)$ satisfying $K^*Kx_\alpha + \alpha D^*Dx_\alpha = K^*g$.

Proof ([19]). A minimum exists at $w \in W$ if and only if

$$\frac{d}{dt} F_\alpha(w + tv)\bigg|_{t=0} = 0$$

for all $v \in W$. This is easily seen to be equivalent to

$$(K^*Ku - K^*g, v) = -\alpha(Du, Dv)$$

for all $v \in W$ and hence $Du \in \mathcal{D}(D^*)$, i.e., $u \in \mathcal{D}(D^*D)$ and

$$K^*Ku + \alpha D^*Du = K^*g. \tag{7}$$

If $N(K) \cap N(D) = \{0\}$ this equation can have at most one solution, for otherwise there would exist a nonzero $u \in \mathcal{D}(D^*D)$ with

$$(K^*Ku + \alpha D^*Du, u) = 0$$

i.e.,

$$\|Ku\|^2 = -\alpha\|Du\|^2 \qquad (\alpha > 0)$$

and hence

$$Ku = Du = 0.$$

We may rewrite (7) as

$$(K^*K - \alpha I)u + \alpha(D^*D + I)u = K^*g .$$

However $D^*D + I : \mathcal{D}(D^*D) \to L^2[0,1]$ has a symmetric compact inverse G(see [28, p.185]) and hence this is equivalent to

$$G(K^*K - \alpha I)u + \alpha u = GK^*g \tag{8}$$

which is a Fredholm equation of the second kind. This equation has a solution (by the Fredholm alternative) when

$$GK^*g \in N((K^*K - \alpha I)G + \alpha I)^\perp.$$

However, $v \in N((K^*K - \alpha I)G + \alpha I)$ is equivalent to

$$0 = (K^*K - \alpha I)w + \alpha(D^*D + I)w = K^*Kw + \alpha D^*Dw$$

and

$$v = (D^*D + I)w.$$

But this implies $w \in N(K) \cap N(D)$ and hence

$$v \in (D^*D + I)(N(K) \cap N(D)),$$

giving $KGv = 0$. But then

$$(GK^*g, v) = (g, KGv) = 0,$$

i.e., a solution to (8) always exists. #

We now present a somewhat different approach, due to Locker and Prenter [19], to the minimization of (6) which dovetails nicely with our development in the previous sections. Note that since N(D) is finite dimensional, there is a constant $\mu > 0$ such that

$$\|Kx\| \geq \mu \|x\|, \quad \text{for} \quad x \in N(D) \tag{9}$$

(see Nashed [25] for the origin of this condition). As usual, we let $(\cdot,\cdot)$ and $\|\cdot\|$ stand for the L^2-inner product and norm, respectively, below. On W we define the inner products

$$[x,y]_I = (x,y) + (Dx, Dy)$$

and

$$[x,y]_K = (Kx, Ky) + (Dx, Dy)$$

and associated norms

$$\|x\|_I^2 = [x,x]_I, \qquad \|x\|_K^2 = [x,x]_k,$$

respectively.

Since D is a closed linear operator on W, W is a Hilbert space under the inner product structure $[\cdot,\cdot]_I$. The next result [18] shows that if $N(K) \cap N(D) = \{0\}$, then W with the inner product structure $[\cdot,\cdot]_k$ is also a Hilbert space. Note that since K is L^2-continuous, convergence in $\|\cdot\|_I$ implies convergence in $\|\cdot\|_K$. Hence there is a constant M such that

$$\|x\|_K \leqslant M\|x\|_I \quad \text{for} \quad x \in W. \tag{10}$$

LEMMA 3.4.2. *Suppose that* W *and* D *are defined as above and* $N(K) \cap N(D) = \{0\}$. *Then* $(W, [\cdot,\cdot]_K)$ *is a Hilbert space and* $\|\cdot\|$ *is equivalent to* $\|\cdot\|_I$.

Proof Suppose $\{x_n\} \subseteq W$ is a Cauchy sequence with respect to the norm $\|\cdot\|_K$. Since N(D) is finite dimensional, each x_n may be uniquely written as

$$x_n = u_n + v_n \in N(D) + W \cap N(D)^{\perp}.$$

By (9), $\{u_n\}$ is an L^2-Cauchy sequence in the finite dimensional space N(D) and hence $u_n \to u \in N(D)$ in the L^2-sense. Then clearly $0 = Du_n \to Du$ and $Ku_n \to Ku$, and hence

$$\|u_n - u\|_K \to 0 \quad \text{as} \quad n \to \infty . \tag{11}$$

Since D is closed, it follows that $W \cap N(D)^{\perp}$ is complete with respect to the norm $\|\cdot\|_I$. Since $R(D) = R(D|W \cap N(D)^{\perp})$ is complete in $L^2[0,1]$ and

$$D : (W \cap N(D)^{\perp}, \|\cdot\|_I) \to (R(D), \|\cdot\|)$$

is continuous and surjective, there is, by the open mapping theorem, a number $m > 0$ such that

$$\|x\|_K^2 \geqslant \|Dx\|^2 \geqslant m\|x\|_I^2 \tag{12}$$

for $x \in W \cap N(D)^{\perp}$. Now, $\{v_n\}$ is $\|\cdot\|_K$- Cauchy in $W \cap N(D)^{\perp}$ and we find from (12) that $\{v_n\}$ is Cauchy with respect to the norm $\|\cdot\|_I$. Therefore there is a $v \in W \cap N(D)^{\perp}$ with

$$\|v_n - v\|_I \to 0 \quad \text{as} \quad n \to \infty .$$

Therefore, $\|v_n - v\|_K \to 0$ by (10). This, combined with (11) gives

$$\|x_n - (u+v)\|_K \to 0 \quad \text{as} \quad n \to \infty$$

i.e., W is complete with respect to the norm $\|\cdot\|_K$. From (10) we conclude that the two norms $\|\cdot\|_K$ and $\|\cdot\|_I$ are equivalent on W. #

The proof of the next lemma is a straightforward verification.

LEMMA 3.4.3. For $z \in W$, $g \in L^2[0,1]$ and $\beta > 0$,

$$\| Kz - g\|^2 + \beta\| z \|_K^2 = (1-\alpha)\{\| Ku - g\|^2 + \alpha\| Du\|^2 + \alpha\| g \|^2\}$$

where $\alpha = \beta/(\beta+1)$ and $u = (\beta+1)z$.

From this we see that minimizing the functional

$$\| Kz - g\|^2 + \beta\| z \|_K^2 \tag{13}$$

over W is equivalent (via the transformations $\beta \to \alpha = \beta/(\beta+1)$ and $z \to u/(\beta+1)$) to minimizing

$$\| Ku - g\|^2 + \alpha\| Du\|^2 \tag{14}$$

over W. Let $\mathcal{K} := K|W$ be the restriction of K to W and denote by $\mathcal{K}^{\#}$ the adjoint of the bounded linear operator $\mathcal{K} : (W, [\cdot,\cdot]_K) \to L^2[0,1]$. Now, from our previous theory we know that the minimizer of (13) is

$$z_\beta = (\mathcal{K}^{\#}\mathcal{K} + \beta I)^{-1}\mathcal{K}^{\#}g$$

and $\| z_\beta - \mathcal{K}^\dagger g\|_K \to 0$ as $\beta \to 0$ if $g \in \mathcal{D}(\mathcal{K}^\dagger)$, where of course $\mathcal{K}^\dagger$ refers to the generalized inverse of $\mathcal{K}$ with respect to the norm $\| \cdot \|_K$. The minimizer of (14) is, according to (3.4.1),

$$x_\alpha = (K^*K + \alpha D^*D)^{-1}K^*g$$

where K^* and D^* refer to L^2-adjoints. From (3.4.3) we have

$$x_\alpha = (1+\beta)z_\beta \to \mathcal{K}^\dagger g \quad \text{as} \quad \alpha \to 0$$

in the norm $\| \cdot \|_K$.

We summarize these results in the following:

THEOREM 3.4.4. Suppose that $N(K) \cap N(D) = \{0\}$ and $g \in \mathcal{D}(\mathcal{K}^\dagger) = R(K|W) + R(K)^\perp$. Then $\| x_\alpha - \mathcal{K}^\dagger g\|_K \to 0$ as $\alpha \to 0$ where $x_\alpha = (K^*K + \alpha D^*D)^{-1}K^*g$ is the unique minimizer of (4) over W.

A few words concerning the vector $\mathcal{K}^\dagger g$ which is the limit of this approximation procedure are in order. Now, $x := \mathcal{K}^\dagger g$ is a least squares solution of $\mathcal{K}u = g$, i.e., $x \in W$ satisfies

$$\| Kx - g\| = \min_{u \in W} \| Ku - g\|$$

$$= \min_{u \in L^2} \| Ku - g\|$$

since K is L^2-continuous and W is dense in $L^2[0,1]$. Therefore x is a standard least squares solution which happens to lie in W. Also, for any other least squares solution $y \in W$ we have $y = x + z$ where $z \in N(K) \cap W$. Therefore, since x is the least squares solution with minimal $\| \cdot \|_K$ norm, we have

$$\| Kx\|^2 + \| Dx\|^2 = \| x \|_K^2 \leq \| Ky\|^2 + \| Dy\|^2$$

$$= \| Kx\|^2 + \| Dy\|^2$$

i.e., $\| Dx\| \leq \| Dy\|$. Therefore $x = K^{\dagger}g$ is the least squares solution for which $\| Dx\|$ is a minimum.

The results in this section show that regularization with differential operators is a special case of the general theory of regularization in Hilbert space as developed earlier. In particular, all of the results of the previous chapter as well as (3.1.1-4) hold where the Hilbert space H_1 is the space W with the inner product $[\cdot,\cdot]_K$. Convergence in this case is with respect to the norm $\| \cdot \|_K$, however by (3.4.2) this is equivalent to the graph norm for D and hence with respect to $\| \cdot \|_K$ implies L^2-convergence.

REFERENCES

1. Anderssen, R.S., de Hoog, F.R. and Lukas, M.A. (Eds.), The Application and Numerical Solution of Integral Equations, Sijthoff & Noordhoff, Alphen aan den Rijn, The Netherlands, 1980.
2. Arcangeli, R., Pseudo-solution de l'equation Ax = y, C.R. Acad. Sci. Paris, Series A, 263(No. 8) (1966), 282-285.
3. Baker, C.T.H., The Numerical Treatment of Integral Equations, Clarendon Press, Oxford, 1977.
4. Bertero, M., Problemi Lineari Non Ben Posti e Metodi di Regolarizzazione, Publicazioni dell'Istituto di Analisi Globale e Applicazioni, Serie "Problemi non ben posti ed inversi," No.4, Consiglio Nazionale delle Ricerche, Florence, 1982.
5. Butler, J.P., Reeds, J.A. and Dawson, S.V., Estimating solutions of first-kind equations with nonnegative constraints and optimal smoothing, SIAM J. Numer. Anal. 18(1981), 381-397.
6. de Hoog, F.R., Review of Fredholm equations of the first kind, in Anderssen et al. [1], pp. 119-134.
7. Franklin, J.N., On Tikhonov's method for ill-posed problems, Math. Comp. 28(1974), 889-907.
8. Groetsch, C.W., Generalized Inverses of Linear Operators: Representation and Approximation, Dekker, New York, 1977.
9. Groetsch, C.W., On a class of regularization methods, Boll. Un. Mat. Ital. 17-B(1980), 1411-1419.
10. Groetsch, C.W., The parameter choice problem in linear regularization: a mathematical introduction, in Nashed [23].
11. Groetsch, C.W., Comments on Morozov's discrepancy principle, in "Improperly Posed Problems and Their Numerical Treatment," (G. Hämmerlin and K.H. Hoffmann, Eds.), Birkhäuser, Basel, 1983.
12. Groetsch, C.W., On the convergence of the method of regularization for equations of the first kind, Numerical Analysis Report No.52, August, 1980, University of Manchester.

13. Groetsch, C.W., On the asymptotic order of accuracy of Tikhonov regularization, J. Optimiz. Th. Appl. 41(1983), to appear.

14. Groetsch, C.W. and King, J.T., The saturation phenomena for Tikhonov regularization, J. Australian Math. Soc. (Series A) 35(1983), 254-262.

15. Guacaneme, J., Ph.D. Thesis - in preparation.

16. Hardy, G.H., Littlewood, J.E. and Pólya, G., Inequalities, Cambridge University Press, Cambridge, 1934.

17. Hilgers, J.W., Non-iterative methods for solving operator equations of the first kind, Mathematics Research Center Technical Summary Report No.1413, January 1974, Madison, Wisconsin.

18. Ivanov, V.K., Approximate solution of operator equations of the first kind, U.S.S.R. Comput. Math. and Math. Phys. 6(No.6) (1966), 197-205.

19. Locker, J. and Prenter, P.M., Regularization with differential operators I: general theory, J. Math. Anal. Appl. 74(1980), 504-529.

20. Lukas, M.A., Regularization, in Anderssen et al. [1], pp. 151-182.

21. Miller, K., Least-squares methods for ill-posed problems with a prescribed bound, SIAM J. Math. Anal. 1(1970), 52-74.

22. Morozov, V.A., On the solution of functional equations by the method of regularization, Soviet Math. Doklady 7(1966), 414-417.

23. Morozov, V.A., The error principle in the solution of operational equations by the regularization method, USSR Comput. Math. and Math. Phys. 8(No.2) (1968), 63-87.

24. Nashed, M.Z. (Ed.), Ill-posed Problems: Theory and Practice, Reidel, Dordrecht, to appear.

25. Nashed, M.Z. (Ed.), Approximate regularized solutions to improperly posed linear integral and operator equations, in "Constructive and Computational Methods for Differential and Integral Equations" (D. Colton and R.P. Gilbert, Eds.), pp. 289-332, Lecture Notes in Mathematics, Vol. 430, Springer-Verlag, Berlin-Heidelberg-New York, 1974.

26. Phillips, D.L., A technique for the numerical solution of certain integral equations of the first kind, J. Asso. Comput. Mach. 9(1962), 84-97.

27. Ribiere, G., Regularisation d'operateurs, Rev. Frans. Inform et Rech. Oper. 1(No.5) (1967), 57-79.

28. Riesz, F. and Sz.-Nagy, B., Functional Analysis, (translated from the second French edition), Ungar, New York, 1955.

29. Schock, E., On the asymptotic rate of convergence of Tikhonov regularization, to appear.

30. Tikhonov, A.N., Solution of incorrectly formulated problems and the regularization method, Soviet Math. Doklady 4(1963), 1035-1038.

31. Tikhonov, A.N. and Arsenin, V.Y., Solutions of Ill-posed Problems, Wiley, New York, 1977. (translated from the Russian).

32. Tippenhauer, U., Regularization of integral equations of the first kind and approximation by Hermite splines, Preprint No.57, Universität Kaiserslautern, February, 1983.

33. Twomey, S., On the numerical solution of Fredholm integral equations by the inversion of the linear system produced by quadrature, J. Asso. Comput. Mach. 10(1963), 97-101.

34. Vinokurov, V.A., Two notes on the choice of regularization parameter, USSR Comput. Math. and Math. Phys. 12(No.2) (1972), 249-253.

4 Finite dimensional approximations

We now reckon with certain approximations to the minimal norm solution which lie in a finite dimensional subspace of the Hilbert space. Such approximations are not quite numerical in the traditional sense, as the convergence will be established in the Hilbert space; however they are effectively computable, as they depend on only finitely many numerical parameters. We begin by considering two natural finite rank approximations to $K^\dagger$. Next we consider regularized Ritz approximations obtained by minimizing the Tikhonov functional over a finite dimensional subspace and relate these to an algorithm of Marti. Finally, we give a brief exposition of the work of Nashed and Wahba on moment discretization and cross-validation.

4.1. Finite rank approximations

A natural way to generate a finite dimensional approximation to the minimal norm least squares solution of the equation

$$Kx = g \tag{1}$$

is to find the minimal norm least squares solution of the equation

$$K_m x = g \tag{2}$$

where $K_m := K|V_m$ is the restriction of K to a finite dimensional subspace V_m of H_1. We shall assume that the finite dimensional subspaces $\{V_m\}$ increase and are eventually dense in H_1, i.e.,

$$V_1 \subseteq V_2 \subseteq \ldots \qquad \text{and} \quad \overline{\bigcup_{m=1}^{\infty} V_m} = H_1,$$

and that (2) has a unique least squares solution, i.e., $N(K) \cap V_m = \{0\}$. We suppose that $g \in \mathcal{D}(K^\dagger)$ and note that since K_m has finite rank, $H_2 = R(K_m) + R(K_m)^\perp$ and hence $g \in \mathcal{D}(K_m^\dagger)$. We of course are interested if

$$K_m^\dagger g \to K^\dagger g \quad \text{as} \quad m \to \infty.$$

Our first result (see [11]) shows that this is equivalent to the uniform boundedness of the operators $\{R_m\}$ defined by

$$R_m := K_m^\dagger Q_m K$$

where Q_m is the orthogonal projector of H_2 onto $K(V_m)$. Note that

$$R_m^2 = K_m^\dagger Q_m K K_m^\dagger Q_m K = K_m^\dagger Q_m Q_m Q_m K = R_m .$$

hence R_m is a (generally nonorthogonal) projection operator.

THEOREM 4.1.1. $K_m^\dagger g \to K^\dagger g$ <u>for each</u> $g \in \mathcal{D}(K^\dagger)$ <u>if and only if</u> $\{\|R_m\|\}$ <u>is bounded.</u>

Proof Note that

$$K_m^\dagger g = K_m^\dagger Q_m g = K_m^\dagger Q_m Q g = K_m Q_m K K^\dagger g = R_m K^\dagger g \tag{3}$$

where Q is the orthogonal projector of H_2 onto $\overline{R(K)}$. If $K_m^\dagger g = R_m K^\dagger g \to K^\dagger g$ for each $g \in \mathcal{D}(K^\dagger)$, then R_m converges pointwise on the Hilbert space $R(K^\dagger) = N(K)^\perp$. Therefore $\|R_m\| = \|R_m|N(K)^\perp\|$ is bounded by the uniform boundedness principle.

Conversely, suppose that $\{\|R_m\|\}$ is bounded and let $z_m \in V_m$ be the best approximation to $K^\dagger g$ in V_m. Since $N(K_m) = \{0\}$, we have

$$R_m z_m = K_m^\dagger Q_m K z_m = K_m^\dagger K_m z_m = z_m .$$

By (3) we then find:

$$\begin{aligned} \|K_m^\dagger g - K^\dagger g\| &\leq \|K_m^\dagger g - z_m\| + \|z_m - K^\dagger g\| \\ &= \|R_m z_m - R_m K^\dagger g\| + \|z_m - K^\dagger g\| \\ &\leq (\|R_m\| + 1)\|z_m - K^\dagger g\| \to 0 \quad \text{as} \quad m \to \infty. \end{aligned}$$ #

Suppose that $\{u_n, v_n, \mu_n\}$ is a singular system for K and that

$$V_m := \text{span}\{v_1, \ldots, v_m\}.$$

Then it is easy to see that

$$K_m^\dagger g = \sum_{n=1}^{m} \mu_n (g, u_n) v_n ,$$

i.e., $K_m^\dagger g$ is the truncated singular function expansion and

$$R_m y = K_m^\dagger \sum_{j=1}^{m} (Ky,u_j)u_j$$

$$= K_m^\dagger \sum_{j=1}^{m} \mu_j^{-1}(y,v_j)u_j$$

$$= \sum_{j=1}^{m} \mu_j^{-1}(y,v_j)K_m^\dagger u_j$$

$$= \sum_{j=1}^{m} (y,v_j)v_j .$$

Therefore R_m is the orthogonal projector of H_1 onto V_m and hence $\|R_m\| = 1$. In this case convergence occurs and

$$K_m^\dagger g \to \sum_{n=1}^{\infty} \mu_n(g,u_n)v_n = K^\dagger g$$

as we have already seen in (1.3.4).

It is not always possible to guarantee convergence, however. Indeed, the following example of Seidman [25] clearly shows the shortcomings of the approach above. Suppose $\{e_n\}$ is an orthonormal basis for the Hilbert space H and let $V_m = \mathrm{span}\{e_1,\ldots,e_m\}$. Let

$$x := \sum_{n=1}^{\infty} n^{-1} e_n \quad \text{and} \quad g := Kx$$

where

$$K \sum_{n=1}^{\infty} a_n e_n := a_1 e_1 + \sum_{n=2}^{\infty} (\alpha_n a_n + a_1/n)e_n$$

and

$$\alpha_n = \begin{cases} n^{-1} , & n \text{ even} \\ n^{-2}(n+1)^{-2}, & n \text{ odd} \end{cases} .$$

Then $K : H \to H$ is an injective compact operator. Therefore

$$K_m^\dagger g =: x_m = \sum_{n=1}^{m} c_n e_n$$

where

$$\frac{\partial}{\partial c_n} \| K(x_m - x) \|^2 = 0, \qquad n = 1,2,\ldots,m.$$

But,

$$\| K(x_m - x) \|^2 = (c_1 - 1)^2 + \sum_{n=2}^{m} (\alpha_n (c_n - n^{-1}) + (c_1 - 1)/n)^2 +$$

$$+ \sum_{n=m+1}^{\infty} (-\alpha_n/n + (c_1 - 1)/n)^2$$

and hence

$$\alpha_n (c_n - 1/n) + (c_1 - 1)/n = 0, \qquad n = 2,\ldots,m$$

and

$$(c_1 - 1) + \sum_{n=m+1}^{\infty} ((c_1 - 1) - \alpha_n)/n^2 = 0.$$

Therefore,

$$c_1 - 1 = \sum_{n=m+1}^{\infty} \alpha_n n^{-2} / \left[1 + \sum_{n=m+1}^{\infty} 1/n^2\right]$$

and

$$c_n - 1/n = (1 - c_1)/(n\alpha_n), \qquad n = 2,\ldots,m.$$

It follows that for m odd

$$\| x_m - x \|^2 = \sum_{n=1}^{m} (c_n - 1/n)^2 + \sum_{n=m+1}^{\infty} 1/n^2$$

$$\geqslant (c_1 - 1)^2 (1 + \sum_{n=2}^{m} \alpha_n^{-2} n^{-2})$$

$$\geqslant (c_1 - 1)^2 / (m^2 \alpha_m^2)$$

$$= m^{-2}\alpha_m^{-2}\left(\sum_{n=m+1}^{\infty} \alpha_n n^{-2}\right)^2 \Big/ \left(1 + \sum_{n=m+1}^{\infty} 1/n^2\right)^2$$

$$\geqslant \frac{36}{\pi^4} m^{-2} \alpha_m^{-2} \alpha_{m+1}^2 (m+1)^{-4}$$

$$\geqslant \frac{36}{\pi^4} m^2/(m+1)^2,$$

and therefore $\{x_m\}$ does not converge to x.

The example above shows that approximating $K^\dagger g$ by $K_m^\dagger g$, where K_m is the finite rank operator obtained by simply restricting K to a finite dimensional subspace, is not always effective. A different approach is to look to the range space itself for finite dimensional subspaces in order to form finite rank approximations to $K^\dagger$. This tack is taken by Seidman [26] and Engl [6].

Suppose that $\{u_1, u_2, u_3, \ldots\} \subseteq \overline{R(K)}$ is linearly independent and $\text{span}\{u_1, u_2, u_3, \ldots\}$ is dense in $\overline{R(K)}$. Define an operator $r_m: H_2 \to \mathbb{R}^m$ by

$$r_m y := \begin{pmatrix} (y, u_1) \\ \\ (y, u_m) \end{pmatrix}.$$

Let $V_m = \text{span}\{K^* u_1, \ldots, K^* u_m\}$ and define $K_m : H_1 \to \mathbb{R}^m$ by

$$K_m := r_m K,$$

i.e., K_m is defined by the commutative diagram

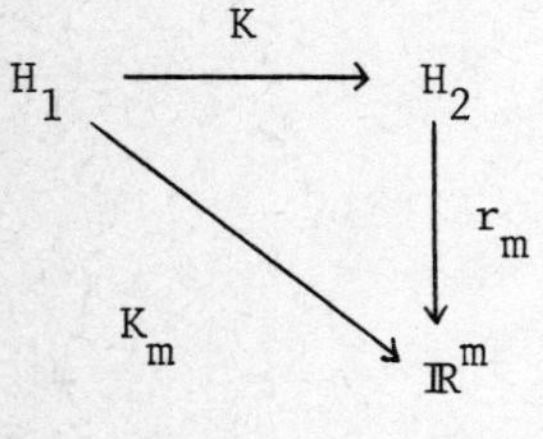

As finite dimensional approximations to $K^\dagger g$ we take

$$x_m := K_m^\dagger r_m g \tag{4}$$

Note that $V_m \subseteq R(K^*) \subseteq N(K)^\perp$. Since $K^\dagger g \in N(K)^\perp = \overline{R(K^*)}$, given $\epsilon > 0$

there is a $z \in N(K^*)^\perp = \overline{R(K)}$ such that

$$\| K^*z - K^\dagger g \| < \epsilon/2.$$

By the assumptions on $\{u_1, u_2, u_3, \ldots\}$ there is an M and a $u \in \mathrm{span}\{u_1, \ldots, u_m\}$, $m \geqslant M$ such that

$$\| u-z \| < \epsilon/(2\| K \|).$$

Therefore

$$\| K^*u - K^\dagger g \| \leqslant \| K^*u - K^*z \| + \| K^*z - K^\dagger g \| < \epsilon.$$

Since $K^*u \in V_m$ for $m \geqslant M$, we see in particular that

$$P_m K^\dagger g \to K^\dagger g \quad \text{as} \quad m \to \infty \tag{5}$$

where P_m is the orthogonal projector of H_1 onto V_m.

If Q denotes the orthogonal projector of H_2 onto $\overline{R(K)}$, then by (4)

$$K_m K^\dagger g = r_m K K^\dagger g = r_m Q g = r_m g$$

since $\{u_i\} \subseteq \overline{R(K)}$. Therefore

$$x_m = K_m^\dagger r_m g = K_m^\dagger K_m K^\dagger g = P_{N(K_m)^\perp} K^\dagger g. \tag{6}$$

However,

$$\begin{aligned} N(K_m)^\perp &= \{z \in H_1 : (z,x) = 0 \text{ if } K_m x = 0\} \\ &= \{z \in H_1 : (z,x) = 0 \text{ if } (x, K^*u_i) = 0,\ i = 1,\ldots,m\} \\ &= \text{bipolar of } \{K^*u_i : i = 1,\ldots,m\} \\ &= \mathrm{span}\{K^*u_i : i = 1,\ldots,m\} = V_m, \text{ i.e.,} \end{aligned}$$

$$P_{N(K_m)^\perp} = P_m. \tag{7}$$

Therefore by (5) and (6), $x_m \to K^\dagger g$ as $m \to \infty$.

Engl [6] has characterized x_m in somewhat different way. Namely, by (6), (7) and (4), we see that x_m is the unique vector satisfying

$$x_m \in V_m \text{ and } (Kx_m, u_i) = (g, u_i),\ i = 1,\ldots,m \tag{8}$$

We may summarize the discussion above in the following theorem.

THEOREM 4.1.2. *Suppose* $g \in \mathcal{D}(K^\dagger)$ *and* $\text{span}\{u_1,u_2,\ldots\}$ *is dense in* $\overline{R(K)}$. *If* x_m *is given by (8) (equivalently (4)), then* $x_m \to K^\dagger g$ *as* $m \to \infty$.

THEOREM 4.1.3. *If* $g \notin \mathcal{D}(K^\dagger)$, *then* $\{x_m\}$ *has no weakly convergent subsequence; in particular,* $\|x_m\| \to \infty$ as $m \to \infty$.

Proof   Note that by (8)

$$0 = (Kx_m - g, u_i) = (Kx_m - Q_m g, u_i), \quad i = 1,2,\ldots,m \tag{9}$$

where Q_m is the orthogonal projector of H_2 onto span $\{u_1,\ldots,u_m\}$. If $\{x_m\}$ has a subsequence, which we again denote by $\{x_m\}$, which is weakly convergent, say

$$x_m \overset{w}{\to} z$$

then since K is compact we have $Kx_m \to Kz$. However, $Q_m g \to Qg$ since $\text{span}\{u_1,u_2,\ldots\}$ is dense in $\overline{R(K)}$. Therefore, by (9), $Kz - Qg \in \overline{R(K)} \cap \overline{R(K)}^\perp$, i.e., the projection of g onto $\overline{R(K)}$ lies in R(K). Therefore

$$g \in R(K) + R(K)^\perp = \mathcal{D}(K^\dagger). \quad \#$$

We also note that if $x = K^\dagger g \in R(K^*)$, say $x = K^* w$, then we have the error estimate

$$\|x - x_m\| = \|(I - P_m)K^* w\| \leq \gamma_m \|w\|,$$

where

$$\gamma_m = \|(I - P_m)K^*\|.$$

We will have more to say about the number γ_m in the next section.

Note that if $\{u_n, v_n; \mu_n\}$ is a singular system for K and $V_m = \text{span}\{K^*u_1,\ldots,K^*u_m\} = \text{span}\{v_1,\ldots,v_m\}$, then by (6) and (1.3.4)

$$x_m = P_m K^\dagger g = \sum_{n=1}^{m} \mu_n (g,u_n) v_n,$$

i.e., in this special case we find again that x_m is the truncated singular function expansion.

We now investigate (following Engl [6]) the case in which the data are in error. Suppose that

$$\| g - g^{\delta} \| \leq \delta$$

and that the u_i are unit vectors. Then

$$\| r_m g - r_m g^{\delta} \|_m \leq \delta$$

where $\| \cdot \|_m$ is the Euclidean norm on $\mathbb{R}^m$. Let

$$x_m^{\delta} = K_m^{\dagger} r_m g^{\delta}.$$

LEMMA 4.1.4. $\| x_m - x_m^{\delta} \|^2 = (A_m^{-1} c^m, c^m)_m$, <u>where</u> $(\cdot,\cdot)_m$ <u>is the Euclidean inner product on</u> $\mathbb{R}^m$, $c^m := r^m(g - g^{\delta})$, <u>and</u> A_m <u>is the m by m matrix</u> $[(K^* u_i, K^* u_j)]$.

Proof By (8) we have

$$x_m = \sum_{i=1}^{m} a_i K^* u_i \quad \text{where } a := \begin{pmatrix} a_1 \\ \vdots \\ a_m \end{pmatrix}$$

satisfies $A_m a = r_m g$. Note that A_m is symmetric and nonsingular since $\{u_i\} \subseteq \overline{R(K)} = N(K^*)^{\perp}$.

Similarly we find that

$$x_m^{\delta} = \sum_{i=1}^{m} a_i^{\delta} K^* u_i, \quad \text{where } A_m a^{\delta} = r_m g^{\delta}$$

and hence

$$\| x_m - x_m^{\delta} \|^2 = \sum_{i=1}^{m} \sum_{j=1}^{m} (a_i - a_i^{\delta})(a_j - a_j^{\delta})(K^* u_i, K^* u_j)$$

$$= (A_m^{-1} r_m(g - g^{\delta}), r_m(g - g^{\delta}))_m. \quad \#$$

It follows that if λ_m is the smallest eigenvalue of A_m, then

$$\| x_m - x_m^{\delta} \|^2 \leq \lambda_m^{-1} \| r_m(g - g^{\delta}) \|^2 \leq \delta^2/\lambda_m. \tag{10}$$

Therefore the number λ_m assumes the role that the regularization parameter played in (2.4.1). Exactly as in (2.4.1) one may now establish the following [6]:

THEOREM 4.1.5. Let λ_m be the smallest eigenvalue of A_m and suppose $\delta_m \to 0$. Let $M := \limsup \delta_m^2/\lambda_m$ and $m := \liminf \delta_m^2/\lambda_m$. Then $\{x_m^\delta\}$ is strongly regular, weakly regular, strongly divergent, or weakly divergent, respectively, according as $M = m = 0$, $M < \infty$, $M = m = \infty$, or $m > 0$, respectively.

From (10) we see that a large value of λ_m will tend to minimize the error in x_m due to noise in the data. In the special case when $\{u_n, v_n; \mu_n\}$ is a singular system for K (with $\mu_1 \leqslant \mu_2 \leqslant \ldots$), then

$$A_m = ((K^*u_i, K^*u_j)) = \operatorname{diag}(\mu_1^{-2}, \ldots, \mu_m^{-2})$$

and hence $\lambda_m = \mu_m^{-2}$. The next result [6] shows that, if the basis functions are orthonormal, then the truncated singular function expansion method is optimal in this sense, i.e., for all choices of orthonormal basis functions $\lambda_m \leqslant \mu_m^{-2}$.

THEOREM 4.1.6. Let $\{\phi_n, \psi_n; \mu_n\}$ be a singular system for K. Among all choices $\{u_1, u_2, \ldots\} \subseteq \overline{R(K)}$ with $(u_i, u_j) = \delta_{ij}$, we have $\lambda_m \leqslant \mu_m^{-2}$.

Proof Let $U_m := \operatorname{span}\{u_1, \ldots, u_m\}$ and suppose $u \in U_m$ is a unit vector satisfying

$$(u, \phi_j) = 0, \quad j = 1, \ldots, m-1$$

(such a u exists since U_m is m-dimensional). Then

$$(\hat{K}u, u) \leqslant \sup\{(\hat{K}y, y) : \|y\| = 1, (y, \phi_j) = 0, j = 1, \ldots, m-1\}$$
$$= \mu_m^{-2},$$

by a familiar variational characterization of the mth eigenvalue of a compact self-adjoint operator (see, e.g., [9, p.154]).

If $u = \sum_{i=1}^{m} \alpha_i u_i$, then $\|\alpha\|_m = 1$ (Euclidean norm) and hence by Rayleigh's principle [9, p.31]

$$(\hat{K}u, u) = \sum_{i,j=1}^{m} \alpha_i\alpha_j (K^*u_i, K^*u_j) = (A_m\alpha, \alpha)_m \geq \lambda_m .$$

Therefore $\lambda_m \leq \mu_m^{-2}$. #

4.2. A regularized Ritz approach

Again we assume that $\{V_m\}$ is an expanding sequence of finite dimensional subspaces of H_1 whose union is dense in H_1. We produce a finite dimensional approximation x_m^α to $K^\dagger g$ by minimizing the Tikhonov functional

$$F_\alpha(z) = \| Kz - g\|^2 + \alpha\| z \|^2 \tag{1}$$

over the finite dimensional space V_m. If $\{v_1, v_2,\ldots,v_{n(m)}\}$ is a basis for V_m, then it is easy to see that x_m^α is given by

$$x_m^\alpha = \sum_{i=1}^{n(m)} y_i v_i$$

where $y \in \mathbb{R}^{n(m)}$ is the unique solution of the n(m) by n(m) linear system

$$(B_m + \alpha M_m)y = w \tag{2}$$

where $w = ((Kv_1, g),\ldots,(Kv_{n(m)}, g))^T$, $M_n = [(v_i, v_j)]$ and $B_m = [(Kv_i, Kv_j)]$. Or equivalently, one could give the coordinate free representation

$$x_m^\alpha = (K_m^*K_m + \alpha I_m)^{-1} K_m^* g = (\tilde{K}_m + \alpha I_m)^{-1} K_m^* Q_m g \tag{3}$$

where $K_m := K|V_m$ is the restriction of K to V_m, I_m is the identity operator on V_m and Q_m is the orthogonal projector of H_2 onto $K(V_m)$.

Of course our aim is to relate α to m, say $\alpha = \alpha_m$, in such a way that

$$x_m^{\alpha_m} \to K^\dagger g \quad \text{as} \quad m \to \infty.$$

The success of this undertaking will depend on how well K_m approximates K, or equivalently how small K is on $V_m^\perp$, that is, how quickly the number

$$\gamma_m := \| K(I - P_m)\| \tag{4}$$

becomes small, where P_m is the orthogonal projector of H_1 onto V_m. Since K is assumed to be compact, the properties of the subspaces $\{V_m\}$ give us the following:

LEMMA 4.2.1. $\gamma_m \to 0$ as $m \to \infty$.

Proof Note that $\gamma_m = \|(I - P_m)K^*\|$. Since the $V_1 \subseteq V_2 \subseteq \ldots$. and the union of the V_m is dense in H_1, the real valued continuous functions

$$f_m(y) := \|(I - P_m)y\|$$

converge pointwise and monotonically to zero. Therefore by Dinis lemma $f_m(y) \to 0$ uniformly on compact subsets of H_1. Since K^* is compact it follows that $f_m(y) \to 0$ uniformly for $y \in K^*B$ where B is the unit ball of H_2, i.e., $\gamma_m \to 0$. #

Using a different method of proof one can in fact show that the lemma holds under less stringent conditions on V_m (see, e.g., [23]).

To illustrate the lemma, let us consider as an example the equation

$$\int_0^1 k(s,t)x(t)dt = g(t).$$

We will denote the L^2-norm on $[0,1]$ by $\|\cdot\|$ and the uniform norm by $\|\cdot\|_\infty$. Suppose that V_m is the space of piecewise cubic splines on $[0,1]$ with a mesh width of $1/m$. Now,

$$\gamma_m = \|K(I - P_m)\| = \|(I - P_m)K^*\|$$

and $$K^*x\,(t) = \int_0^1 k(s,t)x(s)ds.$$

Let $S_m f$ denote the cubic spline interpolating the function f on the mesh. By the variational property of the projector P_m we then have

$$\|(I - P_m)K^*x\| \leqslant \|K^*x - S_m K^*x\| \leqslant \|K^*x - S_m K^*x\|_\infty .$$

Suppose that $\|x\| = 1$ and $k_s(t) = k(s,t)$. We also suppose that

$$\frac{\partial^4}{\partial t^4} k(s,t) \in C([0,1] \times [0,1]).$$

Then

$$|K^*x(t) - S_m K^* x(t)| = |\int_0^1 (k_s(t) - S_m k_s(t))\ x(s)ds|$$

and there is a constant M such that

$$|k_s(t) - S_m k_s(t)| \leq M\ m^{-4}$$

for $s,t \in [0,1]$ (see, e.g., [22, p.112]). By the Cauchy-Schwarz inequality we then have

$$|K^*x(t) - S_m K^* x(t)| \leq M\ m^{-4},$$

i.e., $\| (I - P_m)K^*x\| \leq M\ m^{-4}$ for all x with $\|x\| = 1$, that is $\gamma_m = O(m^{-4})$.

In a similar way one can derive estimates for γ_m using other spaces of splines as well as using higher order regularization in the space H_1 (see [12]).

We now begin a convergence analysis for the approximations $\{x_m^\alpha\}$ obtained by minimizing F_α over V_m. The results will depend on relating α to m, say $\alpha = \alpha_m$, and on the interplay between the numbers α_m and γ_m. In the error-free case, once these numbers are related, the free variable will be m which is a measure of the dimension of V_m and we will investigate the convergence $x_m^\alpha \to K^\dagger g$ as $m \to \infty$ (for related results for a well-posed self-adjoint problem using a different functional, see [24]). For the case in which the data are in error to a level δ we will relate m to δ in such a way that the approximations are regular, i.e., $x_m^\alpha \to K^\dagger g$ as $\delta \to 0$, in both the strong and the weak sense.

Since x_m^α is the unique minimizer of F_α over V_m, it follows easily that x_m^α may be characterized as the unique member of V_m which satisfies

$$(Kx_m^\alpha - g,\ Kv) + \alpha(x_m^\alpha,\ v) = 0 \tag{5}$$

for all $v \in V_m$. Define a new inner product $[\cdot,\cdot]$ on H_1 by

$$[z,y] = (Kz,\ Ky) + \alpha(z,y)$$

and denote the induced norm by $|\cdot|$, i.e.,

$$|z|^2 = [z,z].$$

This inner product is used only in an intermediate stage of the convergence analysis and hence we do not indicate the dependence on α in the notation. We suppose that $g \in \mathcal{D}(K^{\dagger})$ and note that (5) is equivalent to

$$(Kx_m^{\alpha} - Qg, Kv) + \alpha(x_m^{\alpha}, v) = 0$$

where Q is the orthogonal projector of H_2 onto $\overline{R(K)}$. In terms of the inner product $[\cdot,\cdot]$ this becomes

$$[x_m^{\alpha} - x, v] = -\alpha(x,v) \tag{6}$$

for all $v \in V_m$, where $x = K^{\dagger}g$. The infinite dimensional Tikhonov approximation

$$x^{\alpha} := (\tilde{K} + \alpha I)^{-1}K^{*}g \tag{7}$$

of course minimizes F_{α} over the entire space H_1 and hence satisfies

$$[x^{\alpha} - x, v] = -\alpha(x,v)$$

for all $v \in H_1$. Subtracting, we therefore obtain immediately

$$[x_m^{\alpha} - x^{\alpha}, v] = 0 \tag{8}$$

for all $v \in V_m$. If we denote by $\mathcal{P}_m$ the $[\cdot,\cdot]$-orthogonal projector of H_1 onto V_m, then we see that (8) is equivalent to the following characterization of x_m^{α} :

LEMMA 4.2.2. $x_m^{\alpha} = \mathcal{P}_m x^{\alpha}$.

Since $x_m^{\alpha} - \mathcal{P}_m x \in V_m$ and since $I - \mathcal{P}_m$ projects onto the orthogonal (with respect to $[\cdot,\cdot]$) complement of V_m, we may prove the following:

LEMMA 4.2.3. $\| x^{\alpha} - x_m^{\alpha}\|^2 \leq (1 + \gamma_m^2/\alpha) \| (I - P_m)x^{\alpha}\|^2$.

Proof By (4.2.2) and the definition of $\mathcal{P}_m$, we have

$$|x^{\alpha} - x_m^{\alpha}|^2 = |x^{\alpha} - \mathcal{P}_m x^{\alpha}|^2 \leq |x^{\alpha} - P_m x^{\alpha}|^2$$

$$= \| K(I - P_m)^2 x^{\alpha}\|^2 + \alpha\| x^{\alpha} - P_m x^{\alpha}\|^2$$

$$\leq (\gamma_m^2 + \alpha) \| (I - P_m)x^\alpha\|^2,$$

and since $\alpha\|\cdot\|^2 \leq |\cdot|^2$, the result follows. #

Using the inequality

$$\| x_m^\alpha - x\| \leq \| x_m^\alpha - x^\alpha\| + \| x^\alpha - x\|$$

where $x = K^\dagger g$ we may combine (4.2.3) with our results from Chapter 2 to obtain convergence theorems for the approximations $\{x_m^\alpha\}$. We assume that $\alpha = \alpha_m \to 0$ as $m \to \infty$.

From (4.2.3) and (2.1.1) we have:

THEOREM 4.2.4. *If* $\gamma_m = O(\sqrt{\alpha_m})$, *then* $x_m^{\alpha_m} \to K^\dagger g$ *as* $m \to \infty$.

Corollary 3.1.2 gives:

THEOREM 4.2.5. *If* $x \in R(K^*)$ *and* $\gamma_m = O(\sqrt{\alpha_m})$, *then* $\| x_m^{\alpha_m} - x\| = O(\sqrt{\alpha_m})$.

Proof If $x = K^*w$, then by (7)

$$x^\alpha = (\tilde{K} + \alpha I)^{-1}K^*\hat{K}w = K^*(\hat{K} + \alpha I)^{-1}\hat{K}w.$$

But $\| (\hat{K} + \alpha I)^{-1}\hat{K}\| \leq 1$ and hence

$$\| (I - P_m)x^\alpha\| = \| (I - P_m)K^*(\hat{K} + \alpha I)^{-1}\hat{K}w\|$$

$$\leq \gamma_m \|w\|.$$

Therefore, by (4.2.3),

$$\| x^{\alpha_m} - x_m^{\alpha_m} \| \leq \sqrt{1 + \gamma_m^2/\alpha_m}\ \gamma_m\|w\| = O(\sqrt{\alpha_m}),$$

and hence by (3.1.2), $\| x_m^{\alpha_m} - x\| = O(\sqrt{\alpha_m})$. #

One can prove in the same way, using (3.1.1), the following:

THEOREM 4.2.6. *If* $x \in R(K^*K)$ *and* $\gamma_m = O(\alpha_m)$, *then* $\| x_m^{\alpha_m} - x\| = O(\alpha_m)$.

We now take up the question of regularity, that is, suppose that g is not exactly known but we have approximate data $\tilde{g}$ satisfying

$$\| g - \tilde{g}\| \leq \delta.$$

We will denote by $\tilde{x}_m^\alpha$ the minimizer of

$$\| Kz - \tilde{g}\|^2 + \alpha\|z\|^2$$

over V_m. As in (5), we find that this is equivalent to

$$(K\tilde{x}_m^\alpha - \tilde{g}, Kv) + \alpha(\tilde{x}_m^\alpha, v) = 0$$

for all $v \in V_m$, or equivalently

$$[\tilde{x}_m^\alpha - x, v] = -\alpha(x,v) - (g - \tilde{g}, Kv) \tag{9}$$

for all $v \in V_m$. We now establish a stability estimate as in (2.3.2).

LEMMA 2.4.7. $\| x_m^\alpha - \tilde{x}_m^\alpha\| \leq \delta/\sqrt{\alpha}$.

Proof Subtracting (9) from (8), we have

$$[x_m^\alpha - \tilde{x}_m^\alpha, v] = (g - \tilde{g}, Kv) \tag{10}$$

for all $v \in V_m$. Setting $v = x_m^\alpha - \tilde{x}_m^\alpha$ and using $\alpha\|\cdot\|^2 \leq |\cdot|^2$, this gives

$$\alpha\| x_m^\alpha - \tilde{x}_m^\alpha\|^2 \leq |x_m^\alpha - \tilde{x}_m^\alpha|^2 = (g - \tilde{g}, K(x_m^\alpha - \tilde{x}_m^\alpha)).$$

However, $\| g - \tilde{g}\| \leq \delta$ and as in (2.3.1) we find that

$$\| K(x_m^\alpha - \tilde{x}_\alpha^m)\| \leq \delta$$

and the result follows. #

From this lemma we obtain

$$\begin{aligned}\| x - \tilde{x}_m^\alpha\| &\leq \| x - x_m^\alpha\| + \| x_m^\alpha - \tilde{x}_m^\alpha\| \\ &\leq \| x - x_m^\alpha\| + \delta/\sqrt{\alpha}\ .\end{aligned} \tag{11}$$

Suppose now that $m = m_\delta \to \infty$ as $\delta \to 0$. From (11) and (4.2.4) - (4.2.6), respectively, we have:

THEOREM 4.2.8. <u>If</u> $\alpha = \alpha(m_\delta)$ <u>and</u> $\gamma = \gamma(m_\delta)$ <u>satisfy</u> $\gamma = O(\sqrt{\alpha})$ <u>and</u> $\delta = o(\sqrt{\alpha})$, <u>then</u> $\tilde{x}_m^\alpha \to x$ <u>as</u> $\delta \to 0$.

THEOREM 4.2.9. *If* $x \in R(K^*)$, $\alpha = C\delta$ *and* $\gamma = \gamma(m_\delta) = O(\delta^{1/2})$, *then* $\|\tilde{x}_m^\alpha - x\| = O(\delta^{1/2})$.

THEOREM 4.2.10. *If* $x \in R(\tilde{K})$, $\alpha = C\delta^{2/3}$ *and* $\gamma = \gamma(m_\delta) = O(\delta^{2/3})$, *then* $\|\tilde{x}_m^\alpha - x\| = O(\delta^{2/3})$.

For numerical illustrations of related results the reader is referred to [12].

We now take up the question of weak convergence (see [13]).

LEMMA 4.2.11. *If* $\alpha = \alpha(m_\delta)$ *and* $\gamma = \gamma(m_\delta)$ *satisfy* $\gamma = O(\sqrt{\alpha})$ *and* $y \in V_k$ *for some* k, *then*

$$(x_m^\alpha - \tilde{x}_m^\alpha, K^*Ky) \to 0 \quad \text{as} \quad \delta \to 0.$$

Proof Since $V_k \subseteq V_m$ for $m \geqslant k$, we have by (10) and (2.4.7),

$$|(x_m^\alpha - \tilde{x}_m^\alpha, K^*Ky)| \leqslant \alpha|(x_m^\alpha - \tilde{x}_m^\alpha, y)| + |(g - \tilde{g}, Ky)|$$

$$\leqslant \delta(\sqrt{\alpha} + \|K\|)\,\|y\|. \quad \#$$

LEMMA 4.2.12. *If* $\gamma = O(\sqrt{\alpha})$, $\delta = O(\sqrt{\alpha})$ *and* $z \in N(K)$, *then* $(x_m^\alpha - \tilde{x}_m^\alpha, z) \to 0$ *as* $\delta \to 0$.

Proof As noted in (3), x_m^α has the representation

$$x_m^\alpha = (K_m^* K_m + \alpha I)^{-1} K_m^* g$$

and $\tilde{x}_m^\alpha$ has a similar representation with g replaced by $\tilde{g}$. Since

$$(v, K^*y) = (Kv, y) \quad \text{for} \quad v \in H_1 \text{ and } y \in H_2$$

and

$$(v, K_m^*y) = (K_m v, y) \text{ for } \quad v \in V_m \text{ and } y \in H_2$$

we have $(v, K^*y - K_m^*y) = 0$ for all $y \in H_2$ and $v \in V_m$, i.e.,

$$K_m^* = P_m K^*$$

where P_m is the orthogonal projector of H_1 onto V_m. We therefore have

$$(\tilde{x}_m^\alpha - x_m^\alpha, z) = (K_m^*(K_m K_m^* + \alpha I)^{-1}(\tilde{g} - g), z)$$

$$= (P_m K^*(K_m^* K_m + \alpha I)^{-1}(\tilde{g} - g), z)$$

$$= ((I - P_m)K^*(K_m K_m^* + \alpha I)^{-1}(\tilde{g} - g), z)$$

since $z \in N(K)$. But,

$$|((I - P_m)K^*(K_m^* K_m + \alpha I)^{-1}(\tilde{g} - g), z)|$$

$$= |((K_m^* K_m + \alpha I)^{-1}(\tilde{g} - g), K(I - P_m)^2 z)|$$

$$\leq (\delta\gamma/\alpha)\|(I - P_m)z\| \to 0 \text{ as } m = m_\delta \to \infty. \quad \#$$

Consider now a sequence $\{g_n\}$ of increasingly accurate data satisfying

$$\|g - g_n\| \leq \delta_n \to 0 \quad \text{as} \quad n \to \infty.$$

For each n we suppose that the functional (1) is minimized over a subspace V_m where $m = m(\delta_n) \to \infty$ as $n \to \infty$. We suppose that $\alpha = \alpha(m(\delta_n)) \to 0$ and $\gamma = \gamma(m(\delta_n))$ satisfies $\gamma = O(\sqrt{\alpha})$. To simplify notation we will denote the minimum of (1) in V_m by $x_{m(\delta_n)}$ and the minimum of (1) with g replaced by g_n will be designated $x_{m(\delta_n)}$. Our next result shows that if the condition $\delta = o(\sqrt{\alpha})$ in (4.2.8) is replaced by $\delta = O(\sqrt{\alpha})$, then weak convergence obtains.

THEOREM 4.2.13. <u>Using the notational conventions above, if</u> $\delta = O(\sqrt{\alpha})$ <u>then</u> $\tilde{x}_{m(\delta_n)} \overset{w}{\to} x$ <u>as</u> $n \to \infty$.

Proof In light of (4.2.4), it is enough to show that the sequence $\{x_{m(\delta_n)} - \tilde{x}_{m(\delta_n)}\}$ converges weakly to zero as $n \to \infty$.

Since $\delta = O(\sqrt{\alpha})$, we see from (2.4.7) that the sequence is uniformly bounded, say

$$\|x_{m(\delta_n)} - \tilde{x}_{m(\delta_n)}\| \leq M.$$

Therefore by the Banach-Steinhaus theorem it suffices to show that

$$(x_{m(\delta_n)} - \tilde{x}_{m(\delta_n)}, z) \to 0 \quad \text{as} \quad n \to \infty$$

for all z in a dense subspace of H_1. By (4.2.12) this certainly holds for all $z \in N(K) = N(K^*K) = R(K^*K)^{\perp}$.

Suppose then that $z = K^*Kw$. Given $\epsilon > 0$ there is a k and a $y \in V_k$ such that $M\|K\|^2 \|w-y\| < \epsilon$. Therefore

$$|(x_{m(\delta_n)} - \tilde{x}_{m(\delta_n)}, z)| \leq |(x_{m(\delta_n)} - \tilde{x}_{m(\delta_n)}, K^*K(w-y))|$$

$$+ |(x_{m(\delta_n)} - \tilde{x}_{m(\delta_n)}, K^*Ky)|.$$

But $(x_{m(\delta_n)} - \tilde{x}_{m(\delta_n)}, K^*Ky) \to 0$ by (4.2.11). Therefore we find that

$$(x_{m(\delta_n)} - \tilde{x}_{m(\delta_n)}, z) \to 0 \quad \text{as} \quad n \to \infty$$

for each z in the dense subspace $R(K^*K) + R(K^*K)^{\perp}$ of H_1, i.e., $\tilde{x}_{m(\delta_n)} \overset{w}{\to} x$. #

We now see what can be done to place the discrepancy principle in this finite dimensional setting (see [10]). Assume that the available data $\tilde{g}$ satisfies

$$\|g - \tilde{g}\| \leq \delta < c\|\tilde{g}\| \tag{12}$$

where $0 < c < 1$. The last inequality is just another way of saying that the signal to noise ratio is bounded below by a constant $1/c > 1$. Again Q_m will denote the orthogonal projector of H_2 onto $K(V_m)$.

LEMMA 4.2.14. <u>Suppose m is fixed and</u> $\delta/c < \|Q_m\tilde{g}\|$, <u>then there is a unique</u> $\alpha = \alpha(m)$ <u>such that</u> $\|Kx_m^{\alpha} - Q_m\tilde{g}\| = \delta/c$.

Proof Note that $\|Kx_m^{\alpha} - Q_m\tilde{g}\| = \|K_m x_m^{\alpha} - Q_m\tilde{g}\|$, therefore the lemma follows from (3.3.1) by replacing K, g^{δ}, δ in (3.3.1) by K_m, $Q_m\tilde{g}$, and δ/c, respectively. #

Given δ, suppose that we choose $m = m(\delta)$ in such a way that

$$\delta/c < \|Q_m\tilde{g}\| \quad \text{and} \quad \gamma_m = o(\delta) \tag{13}$$

(by (12) and (4.2.1) such can always be done).

LEMMA 4.2.15. If $m = m(\delta)$ is chosen to satisfy (13) and $\alpha = \alpha(m(\delta))$ satisfies (4.2.14), then $\|\tilde{x}_m^\alpha\| \leq \|x\|$ for δ sufficiently small, where $x = K^\dagger g$.

Proof Let P_m be the orthogonal projector of H_1 onto V_m. The inequality $F_\alpha(\tilde{x}_m^\alpha) \leq F_\alpha(P_m x)$ gives

$$\begin{aligned}
&\|K\tilde{x}_m^\alpha - Q_m\tilde{g}\|^2 + \|(I - Q_m)\tilde{g}\|^2 + \alpha\|\tilde{x}_m^\alpha\|^2 \\
&\leq \|KP_m x - \tilde{g}\|^2 + \alpha\|P_m x\|^2 \\
&\leq \|KP_m x - Q_m\tilde{g}\|^2 + \|(I - Q_m)\tilde{g}\|^2 + \alpha\|x\|^2
\end{aligned}$$

and hence

$$\delta^2/c^2 - \|KP_m x - Q_m\tilde{g}\|^2 + \alpha\|\tilde{x}_m^\alpha\|^2 \leq \alpha\|x\|^2 .$$

However, by (12) and (13),

$$\begin{aligned}
\|KP_m x - Q_m\tilde{g}\| &= \|Q_m(KP_m x - \tilde{g})\| \\
&\leq \|K(P_m - I)x\| + \delta = o(\delta) + \delta < \delta/c
\end{aligned}$$

for δ sufficiently small, and the result follows. #

Note that under the conditions of (4.2.15) we have

$$\lim_{\delta\to 0} K\tilde{x}_m^\alpha - g = \lim_{\delta\to 0} K\tilde{x}_m^\alpha - Q_m\tilde{g} = 0.$$

Therefore one can use the same argument as in the proof of (3.3.3) to establish the following:

THEOREM 4.2.16. Under the conditions of (4.2.15), $\tilde{x}_m^\alpha \to x$ as $\delta \to 0$.

We may also obtain an analogue of (3.3.5):

THEOREM 4.2.17. Let $m = m(\delta)$ satisfy (13) and $\alpha = \alpha(m(\delta))$ satisfy (4.2.14). If $x \in R(K^*)$, then $\|\tilde{x}_m^\alpha - x\| = O(\sqrt{\delta})$.

Proof Suppose $x = K^* w$. Then, in view of (4.2.), we find exactly as in the proof of (3.3.5) that

$$\| \tilde{x}_m^\alpha - x \| \leq 2 \| w \| \; \| K\tilde{x}_m^\alpha - g \|$$

for small δ. However,

$$\| K\tilde{x}_m^\alpha - g \| \leq \| K\tilde{x}_m^\alpha - \tilde{g} \| + \delta$$

and

$$\| K\tilde{x}_m^\alpha - \tilde{g} \|^2 = \delta^2/c^2 + \| (I - Q_m)\tilde{g} \|^2 .$$

Moreover,

$$\begin{aligned} \| (I - Q_m)\tilde{g} \| &\leq \delta + \| (I - Q_m) g \| \\ &\leq \delta + \| g - KP_m x \| \\ &\leq \delta + \| K(I - P_m) x \| \leq \delta + \gamma_m \\ &= \delta + o(\delta) \end{aligned}$$

and the result follows. #

4.3. Marti's method

A geometrically motivated algorithm of Marti [16], [17] is closely related to the method studied in the previous section (see [11]). We will study the version of Marti's algorithm as presented in [18].

Suppose that x is the minimum norm solution of

$$Kx = g$$

where $g \in \mathcal{D}(K^\dagger)$ and let $\{V_m\}$, as before, be an expanding sequence of finite dimensional subspaces whose union is dense in H_1. As an approximation to $x = K^\dagger g$ Marti takes the vector x_m of minimum norm in the set

$$V_m \cap [x + c_m K^{-1}(U)]$$

where U is the closed unit ball in H_2 and

$$c_m = (a_m^2 + b_m^2 - \| Qg - g \|^2)^{1/2}$$

where

$$\begin{aligned} a_m &= \inf\{ \| Kv - g \| : \; v \in V_m \} \qquad (1) \\ &= \| (I - Q_m) g \| \end{aligned}$$

and b_m is a sequence of positive numbers converging to zero and satisfying

$$\lim_m \| P_m x - x\| / b_m = 0$$

where P_m is the orthogonal projector of H_1 onto V_m, Q_m is the orthogonal projector of H_2 onto $K(V_m)$ and Q is the orthogonal projector of H_2 onto $\overline{R(K)}$.

Note that

$$a_m^2 = \| (I - Q_m)g\|^2 = \|g\|^2 - \|Q_m g\|^2$$

$$\geq \|g\|^2 - \|Qg\|^2 = \|g - Qg\|^2 \tag{2}$$

and hence c_m is positive. Also, the condition defining x_m is $x_m \in V_m$, has minimum norm and satisfies

$$\| K(x_m - x) \| \leq c_m$$

i.e., $\| Kx_m - Qg\|^2 \leq a_m^2 + b_m^2 - \|g - Qg\|^2$

i.e., $\| Kx_m - g\|^2 \leq a_m^2 + b_m^2$.

As we noted in the discussion prior to the discrepancy principle (Section 3.3), this is equivalent to finding $x_m \in V_m$ with

$$\| x_m \| = \text{minimum}$$

and

$$\| Kx_m - g\|^2 \leq a_m^2 + b_m^2.$$

According to the theory of Lagrange multipliers, this is equivalent to finding $\alpha_m \in \mathbb{R}^+$ and $x_m \in V_m$ with

$$\| Kx_m - g\|^2 = a_m^2 + b_m^2 \tag{3}$$

and $$\alpha_m x_m + K_m^* K_m x_m - K_m^* g = 0 \tag{4}$$

where K_m is the restriction of K to V_m. But note that (4) is the same as (3) of section 4.2 and hence Marti's method is equivalent to the method of Section 4.2 with the parameter α_m chosen by the criterion (3). However, by (1),

$$\| Kx_m - g\|^2 = \| Kx_m - Q_m g\|^2 + \| (I - Q_m)g\|^2$$

$$= \| Kx_m - Q_m g\|^2 + a_m^2$$

and hence (3) is equivalent to

$$\| Kx_m - Q_m g\| = b_m \tag{5}$$

which is strikingly reminiscent of (4.2.14) (with g, b_m playing the roles of $\tilde{g}$, δ/c, respectively). Since $b_m \to 0$, we may show, just as in (4.2.14), that (5) has a unique solution given by (4), at least for m sufficiently large.

LEMMA 4.3.1. <u>For</u> m <u>sufficiently large</u>, $\| x_m\| \leq \|x\|$.

Proof Using the same method as in (4.2.15) we have, in view of (2),

$$\| Kx_m - Q_m g\|^2 + a_m^2 + \alpha_m \| x_m\|^2 \leq \| KP_m x - Qg\|^2 + \| (I - Q)g\|^2 + \alpha_m \| x \|^2$$

$$= \| K(P_m x - x)\|^2 + \| (I - Q)g\|^2 + \alpha_m \| x \|^2$$

$$\leq \|K\|^2 \; \|P_m x - x\|^2 + a_m^2 + \alpha_m \| x \|^2 .$$

Therefore,

$$b_m^2 - \|K\|^2 \| P_m x - x\|^2 + \alpha_m \| x_m\|^2 \leq \alpha_m \| x \|^2 .$$

But, since $\| P_m x - x\| = o(b_m)$, we find that $\| x_m\| \leq \|x\|$, for m sufficiently large. #

By using the same type of argument as in (3.3.3) one can now prove:

THEOREM 4.3.2. $x_m \to K^\dagger g$ <u>as</u> $m \to \infty$ <u>for each</u> $g \in \mathcal{D}(K^\dagger)$.

By following the argument of (4.2.17) one can also prove:

THEOREM 4.3.3. <u>If</u> $x \in R(K^*)$, <u>then</u> $\| x_m - K^\dagger g\| = O(\sqrt{b_m})$.

Moreover, using the argument of (3.3.4) with obvious modifications, the following bound for the parameter α_m in Marti's method can be established (see [11]):

THEOREM 4.3.4. $\alpha_m \leq b_m \|K\|^2/(\|Q_m g\| - b_m)$.

Suppose now that the data consists of a vector $\tilde{g}$ with $\|g - \tilde{g}\| \leq \delta$ and denote by $\tilde{x}_m$ the solution of (3), (4) with g replaced by $\tilde{g}$ (in particular, $a_m = \|(I - Q_m)\tilde{g}\|$, but b_m remains the same). Assume that $m = m(\delta) \to \infty$ as $\delta \to 0$, then we have the following regularity theorem [11]:

THEOREM 4.3.5. <u>If</u> $\delta/b_{m(\delta)} \to 0$ <u>as</u> $\delta \to 0$, <u>then</u> $\tilde{x}_{m(\delta)} \to x$ <u>as</u> $\delta \to 0$.

Proof Let $m = m(\delta)$. As in the proof of 4.3.1 we have

$$b_m^2 + \alpha_m \|\tilde{x}_m\|^2 \leq \|KP_m x - Q_m\tilde{g}\|^2 + \alpha_m \|P_m x\|^2$$

and hence

$$b_m^2 - \|KP_m x - Q_m\tilde{g}\|^2 + \alpha_m \|\tilde{x}_m\|^2 \leq \alpha_m \|x\|^2 .$$

But

$$KP_m x - Q_m\tilde{g} = Q_m(KP_m x - g + g - \tilde{g})$$

and hence

$$b_m - \|KP_m x - Q_m\tilde{g}\| \geq b_m - \|K\| \, \|P_m x - x\| - \delta$$

$$= b_m - o(b_m) - \delta \geq 0$$

for δ sufficiently small. Therefore, for such δ, $\|\tilde{x}_{m(\delta)}\| \leq \|x\|$ and the rest of the proof follows as before. #

4.4. Moment discretization and cross validation.

Consider the integral equation

$$\int_0^1 k(s,t)\, x(t)dt = g(s) \qquad 0 \leq s \leq 1 \tag{1}$$

where $k(\cdot,\cdot)$ is a given continuous kernel and g is a given continuous function. A natural way to approximate a solution of (1) is by the method of collocation or *moment discretization*. That is, given a finite set of points

$$0 \leq s_1 < s_2 < s_3 < \ldots < s_m \leq 1,$$

one seeks a function $x(t)$ which satisfies the m equations

$$\int_0^1 k(s_i, t)x(t)dt = g(s_i), \quad i = 1,\ldots,m.$$

Note that this is equivalent to the

$$\begin{pmatrix} (x, k_{s_1}) \\ \vdots \\ (x, k_{s_m}) \end{pmatrix} = \begin{pmatrix} g_1 \\ \vdots \\ g_m \end{pmatrix} \tag{2}$$

where $k_s(t) := k(s,t)$ and $g_i = g(s_i)$, $i = 1,\ldots,m$.

In this section we investigate some work of Nashed and Wahba on the moment discretization method by relating it to one of the finite rank methods studied in Section 4.1. In order to do this it is necessary to develop the appropriate Hilbert space formalism. To cast (2) in the context of Section 4.1, it is necessary to have a Hilbert space in which

$$g(s_i) = (g, u_i) \quad i = 1,\ldots,m \tag{3}$$

for some vectors $u_i \in R(K)$. It follows from the Riesz theorem that in this Hilbert space the evaluation functionals

$$g \to g(s_i)$$

must be continuous. We now develop some basic properties of such spaces (see [2] and [27] for more details).

A Hilbert space H of real valued functions on $[0,1]$ with inner product $[\cdot,\cdot]$ is called a *reproducing kernel Hilbert space* if the evaluation functionals $f \to f(s)$ are continuous for each $s \in [0,1]$. If H is a reproducing kernel Hilbert space, then by the Riesz theorem there is for each $s \in [0,1]$ a function $\rho_s \in H$ such that

$$f(s) = [f, \rho_s] \tag{4}$$

for each $f \in H$. For this reason the function $\rho(s,t) := \rho_s(t)$ is called the *reproducing kernel* for H. Note that by (4) we have

$$\rho(s,t) = [\rho_s, \rho_t] \quad \text{for } s,t \in [0,1]. \tag{5}$$

Let $\|\cdot\|_\rho$ be the norm on H induced by $[\cdot,\cdot]$, i.e.,

$$\|f\|_\rho^2 = [f,f].$$

If $f_n \to f$ in H, then for $s \in [0,1]$

$$|f_n(s) - f(s)| = |[f - f_n, \rho_s]| \leq \|f - f_n\|_\rho \, \|\rho_s\|_\rho$$

and hence convergence in H implies pointwise convergence. Moreover, by (5),

$$\|\rho_s\|_\rho = \sqrt{\rho(s,s)}$$

and hence if $\rho(s,s)$ is continuous, then convergence in H implies uniform convergence.

In fact, the reproducing quality of the kernel ρ (i.e., (4)) shows that weak convergence in H implies pointwise convergence. Indeed, if $f_n \overset{w}{\to} f$ in H, then

$$f_n(s) = [f_n, \rho_s] \to [f, \rho_s] = f(s) \quad \text{for each} \quad s \in [0,1].$$

Note that $L^2[0,1]$ is not a reproducing kernel Hilbert space since L^2-convergence does not imply pointwise convergence. We now give a simple example of a reproducing kernel Hilbert space.

Let

$$H = \{f:[0,1] \to \mathbb{R} : f \text{ is absolutely continuous, } f(0) = 0\},$$

and $\quad [f,g] = \int_0^1 f'(t)g'(t)dt, \quad$ for $f, g \in H$.

Note that $[\cdot,\cdot]$ is in fact an inner product on H. Let $\{g_n\}$ be a Cauchy sequence in H. Then $\{g_n'\}$ is a Cauchy sequence in $L^2[0,1]$ and hence there is an $h \in L^2[0,1]$ with $g_n \to h$ in L^2 as $n \to \infty$. Let

$$g(t) = \int_0^t h(u)du \in H.$$

We then have

$$[g_n - g, g_n - g] = \int_0^1 (g_n'(u) - h(u))^2 du \to 0 \quad \text{as} \quad n \to \infty,$$

i.e., $g_n \to g$ in H. Therefore $(H, [\cdot,\cdot])$ is a Hilbert space. Moreover, if

$$\rho(s,t) = \min(s,t) \quad \text{for} \quad s,t \in [0,1],$$

then for $f \in H$ we have

$$[f, \rho_s] = \int_0^s f'(u)du = f(s) \quad \text{for} \quad s \in [0,1],$$

and hence H is a reproducing kernel Hilbert space. The fact that H, i.e., the range of the integral operator $K : L^2[0,1] \to L^2[0,1]$ defined by

$$Kf(t) = \int_0^t f(u)du,$$

is a reproducing kernel Hilbert space is no accident; we now show that such is always the case (see [21]).

Consider a linear integral operator $K : L^2[0,1] \to L^2[0,1]$ generated by a kernel $k(\cdot,\cdot)$. Define an inner product $[\cdot,\cdot]$ on $R(K)$ by

$$[f,g] = (K^\dagger f, K^\dagger g) \tag{6}$$

where $(\cdot,\cdot)$ is the L^2-inner product. Note that this is indeed an inner product, for if $g \in R(K)$, say $g = Kx$ where $x \in N(K)^\perp$, then if $[g,g] = 0$ we have

$$0 = [g,g] = (K^\dagger g, K^\dagger g) = (x,x)$$

and hence $x = 0$, i.e., $g = Kx = 0$.

Also, $R(K)$ is a Hilbert space under the inner product (6), for if $\{f_n\} \subseteq R(K)$ is Cauchy and $f_n = Kx_n$, $x_n \in N(K)^\perp$, then $\{x_n\}$ is Cauchy in $L^2[0,1]$. Therefore there is an $x \in N(K)^\perp$ with $x_n \to x$. We then have

$$[f_n - Kx, f_n - Kx] = \| x_n - x \| \to 0$$

as $m \to \infty$, where $\|\cdot\|$ is the L^2-norm. Therefore $R(K)$ is complete with respect to the inner product (6). Now let

$$\rho(s,t) = \int_0^1 k(s,u)k(t,u)du \tag{7}$$

and $\rho_s(t) = \rho(s,t)$.

Note that

$$\rho_s(t) = (Kk_s)(t)$$

where $k_s(u) = k(s,u)$. Therefore if P is the L^2-orthogonal projector of $L^2[0,1]$ onto $N(K)^\perp$, we have for $f = Kx$ and $x \in N(K)^\perp$,

$$[f, \rho_s] = (K^\dagger f, K^\dagger \rho_s) = (x, K^\dagger \rho_s)$$

$$= (x, K^\dagger K\, k_s) = (x, Pk_s)$$

$$= (Px, k_s) = (x, k_s)$$

$$= \int_0^1 k(s,u)x(u)du = f(s)$$

and hence $(R(K), [\cdot,\cdot])$ is a reproducing kernel Hilbert space with reproducing kernel given by (7).

For $g \in R(K)$, we then see that (3) holds for the inner product $[\cdot,\cdot]$ and $u_i = \rho_{s_i}$, i.e.

$$g(s_i) = [g, \rho_{s_i}], \quad i = 1,\ldots,m.$$

In order to apply the results of Section 4.1, we only require that the $\{\rho_{s_i}\}$ are dense in the reproducing kernel Hilbert space $R(K)$.

Let

$$\Delta_m = \sup_{s\in[0,1]} \inf\{|s - s_i| : i = 1,2,\ldots,m\}. \tag{8}$$

If k is continuous, then so is each $g \in R(K)$. Therefore if $f \in R(K)$ is orthogonal to each ρ_{s_i}, then

$$f(s_i) = [f, \rho_{s_i}] = 0 \qquad i = 1,2,\ldots$$

and hence if $\Delta_m \to 0$ as $m \to \infty$ it follows that $f = 0$. Under these conditions $\{\rho_{s_i}\}$ is dense in $R(K)$ and we may apply (4.1.2) to obtain [21]:

THEOREM 4.4.1. _Suppose that_ $k(\cdot,\cdot)$ _is continuous and_ $g \in R(K)$. _If_ $\Delta_m \to 0$ _as_ $m \to \infty$, _then_ $\|x_m - K^\dagger g\| \to 0$ _as_ $m \to \infty$.

Of course by x_m above we mean the unique element x of $L^2[0,1]$ satisfying

$$(Kx)(s_i) = [g, \rho_{s_i}] \qquad i = 1,\ldots,m. \tag{9}$$

But note that

$$(Kx)(s_i) = [Kx, \rho_{s_i}] = (x, K^*\rho_{s_i})$$

where K^* is the adjoint of the operator

$$K : (L^2[0,1], (\cdot,\cdot)) \to (R(K), [\cdot,\cdot]).$$

Therefore (9) is equivalent to (8) of Section 4.1, using the inner product $[\cdot,\cdot]$ on $R(K)$ and $u_i = \rho_{s_i}$.

Corresponding to (4.1.3) we have (see [21]):

THEOREM 4.4.2. <u>If</u> $\{x_m\}$ <u>has a weakly convergent subsequence, then</u> $g(s_i) = y(s_i)$ <u>for</u> $i = 1,2,\ldots,$ <u>for some</u> $y \in R(K)$. <u>Consequently, if g does not agree with some</u> $y \in R(K)$ <u>at the points</u> $s_1, s_2, \ldots,$ <u>then</u> $\| x_m \| \to \infty$.

Proof Suppose that $\{x_m\}$ has a weakly convergent subsequence, which we may also denote by $\{x_m\}$, say $x_m \overset{w}{\to} z$ in $L^2[0,1]$. Let $w_m = Kx_m$, then $w_m(s_i) = g(s_i)$, $i = 1,\ldots,m$. We then have, for each $u \in L^2[0,1]$,

$$[w_m - Kz, Ku] = [Kx_m - Kz, Ku]$$

$$= (x_m - z, u) \to 0 \quad \text{as} \quad m \to \infty$$

and hence w_m converges weakly with respect to $[\cdot,\cdot]$ to Kz. But weak convergence with respect to $[\cdot,\cdot]$ implies pointwise convergence and hence

$$g(s_i) = w_m(s_i) \to Kz(s_i) \text{ as } m \to \infty, \quad i = 1,2,\ldots,$$

i.e., $g(s_i) = (Kz)(s_i)$, $i = 1,2,\ldots$. #

In the particular case when $K^\dagger g \in R(K^*)$ and the reproducing kernel satisfies certain smoothness conditions, Wahba (see [28], [29]) has established an asymptotic convergence rate for $\| x_m - K^\dagger g \|$. Namely, if

$$\frac{\partial^j}{\partial s^j} \rho(s,t) \quad \text{is continuous for } s \neq t,\ j = 0,1,2,\ldots,2p$$

and

$$\frac{\partial^j}{\partial s^j}\rho(s,t) \quad \text{is continuous for } s,t \in [0,1], \quad j = 0,1,2,\ldots,2p-2$$

and

$$\lim_{s\uparrow t} \frac{\partial^{2p-1}}{\partial s^{2p-1}}\rho(s,t) \quad \text{and} \quad \lim_{s\downarrow t} \frac{\partial^{2p-1}}{\partial s^{2p-1}}\rho(s,t)$$

exist and are bounded for each $t \in [0,1]$, then $\| x_m - K^\dagger g\| = O(\Delta_m^p)$, where Δ_m is given by (8). We refer the reader to [28] for details.

Consider now the general case $g \in \mathcal{D}(K^\dagger) = R(K) + R(K)^\perp$ (orthogonal complement in $L^2[0,1]$. By the characterization of $K^\dagger g$ as the unique solution x of $K^*Kx = K^*g$ lying in $N(K)^\perp = N(K^*K)^\perp$, we see that

$$K^\dagger g = (K^*K)^\dagger K^*g.$$

Now $g \in R(K) + R(K)^\perp = R(K) + N(K^*)$ implies that $K^*g \in R(K^*K)$. Therefore, replacing K, g in the discussion above by K^*K, K^*g, respectively, we can, by considering $R(K^*K)$ as a reproducing kernel Hilbert space, produce approximations x_m which converge as $m \to \infty$ to $K^\dagger g = (K^*K)^\dagger K^*g$ for any $g \in \mathcal{D}(K^\dagger)$ (see [21], Theorem 4.1).

Nashed and Wahba [20] have also studied Tikhonov regularization in the context of reproducing kernel Hilbert spaces, that is, for operators $K : H_1 \to H_2$ where H_1 and H_2 are reproducing kernel Hilbert spaces. Under appropriate conditions they obtain for the case of imprecise data bounds of the type $O(\sqrt{\delta})$ and $O(\delta^{2/3})$ corresponding to (3.1.3) and (3.1.4). Of course the advantage of the reproducing kernel Hilbert space setting is that the bounds hold in the norm of H_1 and in particular convergence in H_1 implies pointwise convergence.

Finally we consider briefly Wahba's cross validation method for choosing the regularization parameter. Suppose that instead of the equations

$$\int_0^1 k(s_i, t)x(t)dt = g(s_i), \quad i = 1,2,\ldots,m$$

we have

$$\int_0^1 k(s_i, t)x(t)dt = d_i, \qquad i = 1,2,\ldots,m \tag{10}$$

where $d_i = g(s_i) + \epsilon_i$, the numbers ϵ_i representing errors in the measured data values $\{g(s_i)\}$. One might try to regularize (10) by minimizing the expression

$$\frac{1}{m}\sum_{i=1}^{m} (Kx(s_i) - d_i)^2 + \alpha\|x\|^2$$

where $\|\cdot\|$ is the norm on some appropriate Hilbert space, e.g. $L^2[0,1]$ or a reproducing kernel Hilbert space. Note that if $K_m: H_1 \to \mathbb{R}^m$ is defined by

$$(K_m x)_i = Kx(s_i), \quad i = 1,\ldots,m$$

and the norm on $\mathbb{R}^m$ is given by

$$\|z\|_m = \{\sum_{i=1}^{m} z_i^2/m\}^{1/2}$$

then this is just the familiar Tikhonov functional

$$F_\alpha(x) = \|K_m x - d\|_m^2 + \alpha\|x\|^2 .$$

Previously *a priori* choices of α and a choice which depends heavily only on the size of the errors (the discrepancy method) have been considered. However, numerical experiments (see [15]) indicate that to be truly effective the choice of α must depend on the actual errors rather than on just some measure of their overall size. Wahba's cross validation method is a technique for incorporating the actual data points in a scheme to choose a "good" regularization parameter.

Suppose that $K : H_1 \to L^2[0,1]$ is a bounded linear operator generated by a continuous kernel $k(\cdot,\cdot)$ and H_1 is either $L^2[0,1]$ or a reproducing kernel Hilbert space $\mathcal{H}$ with reproducing kernel $\rho(\cdot,\cdot)$ and inner product $[\cdot,\cdot]$. Wahba's theory depends on properties of the kernel

$$Q(s,t) = \begin{cases} \int_0^1 k(s,u)k(t,u)\,du & \text{if } H_1 = L^2[0,1] \\ \int_0^1\int_0^1 k(s,u)k(t,v)\rho(u,v)\,dv\,du & \text{if } H_1 = \mathcal{H}. \end{cases}$$

If $Q(s,t)$ has a Hilbert-Schmidt expansion

$$Q(s,t) = \sum_n \lambda_n u_n(s)u_n(t),$$

then Wahba calls g "very smooth" if

$$\sum_n \lambda_n^{-2} (g, u_n)^2 < \infty \tag{12}$$

where $(\cdot,\cdot)$ is the L^2-inner product.

In the case where $H_1 = L^2[0,1]$, note that Q is the kernel which generates the operator

$$KK^* : L^2[0,1] \to L^2[0,1]$$

where K^* refers to the adjoint of $K : L^2[0,1] \to L^2[0,1]$. If P denotes the orthogonal projector of $L^2[0,1]$ onto $\overline{R(KK^*)} = N(KK^*)^\perp$, then since $u_n \in N(KK^*)^\perp$, (12) is equivalent to

$$\sum_n \lambda_n^{-2} (Pg, u_n)^2 < \infty. \tag{13}$$

However, since $\{u_n; \lambda_n\}$ is an eigensystem for KK^*, $\{u_n, u_n; \lambda_n^{-1}\}$ is a singular system for KK^*. But $Pg \in N(KK^*)^\perp$ and hence (13) is equivalent (by Picard's theorem (1.2.6)) to

$$Pg \in R(KK^*), \quad \text{i.e.,} \quad g \in \mathcal{D}((KK^*)^\dagger).$$

Therefore in the case $H_1 = L^2[0,1]$ Wahba's condition that g is very smooth is equivalent to $g \in \mathcal{D}((KK^*)^\dagger)$.

Consider now the case where $H_1 = \mathcal{H}$, a reproducing kernel Hilbert space with reproducing kernel $\rho(\cdot,\cdot)$ and inner product $[\cdot,\cdot]$. Note that if ϕ is any continuous linear functional on $\mathcal{H}$, then by the Riesz theorem there is a unique $\eta \in \mathcal{H}$ with

$$\phi(f) = [f, \eta], \quad \text{for all } f \in \mathcal{H}.$$

However, by the reproducing quality of the kernel

$$\phi(\rho_s) = [\rho_s, \eta] = \eta(s) \tag{14}$$

i.e., the Riesz representer of the functional ϕ is obtained by applying ϕ to the reproducing kernel. Consider now the operator $K : \mathcal{H} \to L^2[0,1]$ and let $K^*: L^2[0,1] \to \mathcal{H}$ be its adjoint. We then have for $f \in \mathcal{H}$, $g \in L^2[0,1]$:

$$[K^*g, f] = (g, Kf) = \int_0^1 Kf\ (t)\ g(t)dt. \tag{15}$$

We assume with Wahba [29] that the functionals $f \to (Kf)(t)$ are continuous and hence there is a unique $\eta_t \in H$ with

$$(Kf)(t) = [\eta_t, f]$$

where by (14)

$$\eta_t(u) = (K\rho_u)(t) = \int_0^1 k(t,v)\rho(u,v)dv. \tag{16}$$

We therefore have by (15)

$$[K^*g, f] = \int_0^1 [\eta_t, f]\ g(t)\ dt.$$

Consider the continuous linear functional

$$\phi(f) = (Kf, g) = \int_0^1 (Kf)(t)g(t)dt$$

$$= \int_0^1 [\eta_t, f]g(t)dt.$$

By (14),

$$\phi(f) = [h, f]$$

where by (16)

$$h(u) = \phi(\rho_u) = \int_0^1 [\eta_t, \rho_u]g(t)dt = \int_0^1 \eta_t(u)g(t)dt$$

i.e.,

$$[K^*g, f] = [h, f], \quad \text{for all } f \in H.$$

In other words,

$$K^*g(u) = \int_0^1 \eta_t(u)g(t)dt = \int_0^1 \int_0^1 k(t,v)\rho(u,v)dv\ g(t)dt$$

and hence

$$KK^*g(s) = \int_0^1\int_0^1\int_0^1 k(s,u)k(t,v)\rho(u,v)\,dvdu\; g(t)dt$$

$$= \int_0^1 Q(s,t)\; g(t)dt.$$

Therefore in the case $H_1 = \mathcal{H}$ we also see that Q is the kernel of the integral operator

$$KK^* : L^2[0,1] \to L^2[0,1].$$

As above, we again find the following characterization of very smooth:

"g is very smooth if and only if $g \in \mathcal{D}((KK^*)^\dagger)$."

The only difference between the cases $H_1 = L^2[0,1]$ and $H_1 = \mathcal{H}$ is the interpretation of the adjoint.

The idea behind cross validation is to allow the data points themselves to predict a good value of the regularization parameter and moreover that a good value of the parameter should predict missing data points.

Let $x_{\alpha,k}$ be the minimizer of the functional

$$\frac{1}{m}\sum_{\substack{i=1\\ i\neq k}}^{m} (Kx(s_i) - d_i)^2 + \alpha\|x\|^2$$

i.e., (11) with the kth data point omitted. The value of α will then be chosen as $\hat{\alpha}$ where $Kx_{\hat{\alpha},k}$ is nearer to d_k on the average than $Kx_{\alpha,k}$ for any other value of α. Therefore $\hat{\alpha}$ is chosen for its ability to predict any missing data point, given the other data points. Specifically, let

$$V(\alpha) = \frac{1}{m}\sum_{k=1}^{m} (Kx_{\alpha,k}(s_k) - d_k)^2\; w_k(\alpha)$$

where the $w_k(\alpha)$ are certain weights. We would then like to choose $\hat{\alpha}$ so that

$$V(\hat{\alpha}) = \min_{\alpha>0}\; V(\alpha).$$

Of course for $\hat{\alpha}$ to be a good choice for α it should be close to α^*, the minimizer of

$$T(\alpha) = \frac{1}{m} \sum_{i=1}^{m} (Kx_\alpha(s_i) - g(s_i))^2.$$

But note that α^* is not computable since the true values $g(s_i)$ are not known.

Assuming that the errors ϵ_i are uncorrelated, have mean zero and common variance and that $s_i = i/m$ and g is very smooth, Wahba [29] shows that for a certain choice of weights

$$\alpha^* = \hat{\alpha}(1 + o(1))$$

where $o(1) \to 0$ as $m \to \infty$, and where α^* and $\hat{\alpha}$ are the minimizers of the expected values $ET(\alpha)$ and $EV(\alpha)$, respectively. Moreover, $E\|K^\dagger g - x_{\alpha^*}\| \to 0$ as $m \to \infty$.

It would seem that the major drawback of this method of choosing α is the expense of the computations, but see [4] for suggestions for streamlining the calculations.

REFERENCES

1. Anderssen, R.S., de Hoog, F.R., and Lukas, M.A., (Eds). The Application and Numerical Solution of Integral Equations, Sijthoff & Noordhoff, Alphen aan den Rijn, The Netherlands, 1980.
2. Aronszajn, N., Theory of reproducing kernels. Trans. Amer. Math. Soc. 68(1950), 337-404.
3. Baker, C.T.H. and Miller, G.F., (Eds), Treatment of Integral Equations by Numerical Methods, Academic Press, London, 1982.
4. Bates, D.M. and Wahba, G., Computational methods for generalized cross validation with large data sets, in [3].
5. Deuflhard, P. and Hairer, E., (Eds), Inverse Problems in Differential and Integral Equations, Birkhäuser, Boston, 1983.
6. Engl, H.W., On the convergence of regularization methods for ill-posed linear operator equations, in [14].
7. Engl, H.W., Regularization and least squares collocation, in [5].
8. Golberg, M.A., (Ed.), Solution Methods for Integral Equations, Plenum Press, New York, 1978.
9. Gould, S.H., Variational Methods for Eigenvalue Problems, Mathematical Expositions No.10, University of Toronto Press, Toronto, 1957.
10. Groetsch, C.W., Comments on Morozov's discrepancy principle, in [14].
11. Groetsch, C.W., On a regularization - Ritz method for Fredholm equations of the first kind, J. Integral Equations 4(1982), 173-182.
12. Groetsch, C.W., King, J.T., and Murio, D., Asymptotic analysis of a finite element method for Fredholm equations of the first kind, in [3].
13. Groetsch, C.W., and Guacaneme, J., Regularized Ritz approximations for Fredholm equations of the first kind, to appear.
14. Hämmerlin, G. and Hoffmann, K.H., (Eds.), Improperly Posed Problems and Their Numerical Treatment, ISNM, Birkhäuser, Basel, 1983.
15. de Hoog, F.R., Review of Fredholm equations of the first kind, in [1].

16. Marti, J.T., An algorithm for computing minimum norm solutions of Fredholm integral equations of the first kind, SIAM J. Numer. Anal. 15(1978), 1071-1076.

17. Marti, J.T., On the convergence of an algorithm for computing minimum norm solutions of ill-posed problems, Math. Comp. 34(1980), 521-527.

18. Marti, J.T., On a regularization method for Fredholm equations of the first kind using Sobolev spaces, in [3].

19. Nashed, M.Z., On moment discretization and least squares solutions of linear integral equations of the first kind, J. Math. Anal. Appl. 53(1976), 359-366.

20. Nashed, M.Z. and Wahba, G., Approximate regularized solutions to linear operator equations when the data-vector is not in the range of the operator, MRC Technical Summary Report 1265, Madison, Wisconsin, 1973.

21. Nashed, M.Z. and Wahba, G., Convergence rates of approximate least squares solutions of linear integral and operator equations of the first kind, Math. Comp. 28(1974), 69-80.

22. Prenter, P., Splines and Variational Methods, Wiley, New York, 1975.

23. Schock, E., Numerische Lösung Fredholmscher Integralgleichungen, Kaiserslautern, 1982.

24. Schock, E., Regularisierungsverfahren für Gleichungen erster Art mit positiv definiten Operatoren, in [14].

25. Seidman, T.I., Nonconvergence results for the application of least-squares estimation to ill-posed problems, J. Optimiz. Th. Appl. 30(1980), 535-547.

26. Seidman, T.I., Convergent approximation schemes for ill-posed problems, Proc. Conf. on Information Science and Systems, Johns Hopkins University, Baltimore, 1976, pp. 258-262.

27. Shapiro, H.L., Topics in Approximation Theory, Lecture Notes in Mathematics, Vol.187, Springer-Verlag, New York - Heidelberg - Berlin, 1970.

28. Wahba, G., Convergence rates for certain approximate solutions to Fredholm integral equations of the first kind, J. Approximation Theory 7(1973), 167-185.

29. Wahba, G., Practical approximate solutions to linear operator equations when the data are noisy, SIAM J. Numer. Anal. 14(1977), 651-667.

30. Wahba, G., Smoothing and ill-posed problems, in [8].

List of symbols

Subject index

DATE DUE

BOOKS ARE SUBJECT TO
RECALL AFTER TWO WEEKS